GEOGRAPHY: THEORY IN PRACTICE
BOOK ONE
SETTLEMENTS

GEOGRAPHY: THEORY IN PRACTICE
BOOK ONE
SETTLEMENTS

IAIN R. MEYER & RICHARD J. HUGGETT.

Harper & Row, Publishers

London   New York   Hagerstown   San Francisco   Sydney

Harper & Row Ltd
28 Tavistock Street
London WC2E 7PN

British Library Cataloguing in Publication Data
Geography.
   Book 1: Settlements
   1. Geography
   I Huggett, Richard   II. Meyer, Iain
   910   G116
ISBN 0 06 318096 0

Designed by Richard Dewing 'Millions', London
Typeset by Preface Ltd, Salisbury
Printed and bound by A. Wheaton & Co. Ltd, Exeter

Contents

Preface

This book outlines and illustrates the chief theories and concepts of the geography of settlements. These theories and concepts are explained concisely and, as far as possible, in jargon-free language. By means of carefully designed exercises, students should see how these theories work in practice. The exercises are in the main based on original data from extensive field-work or from official sources. Worked examples of the techniques used are included, and carefully posed questions encourage students to interpret and to criticize their results. Where appropriate, sectional paper is added, on which is indicated the size and type of graph required in answer to a question. At the end of each chapter is a list of books, which might be helpful in answering the questions and in essay work. Though written as a course-book, *Settlements* has plenty of scope for teachers to bring in their own examples and sets of data.

Chapter One opens with an explanation of settlement patterns made from the standpoint of economics and is followed by a look at the influence of physical factors. Urban fields are examined in Chapter Two where students are shown how they can be delimited by both field survey methods and mathematical techniques. The structure or morphology of settlements, as explained in models of urban land use, is discussed in Chapter Three. The final chapter studies the way in which climate and streamflow are modified by settlements.

Iain Meyer   **Richard Huggett**
Elstree   Macclesfield

September 1978

CHAPTER ONE
SETTLEMENT PATTERNS

Introduction

Man is by nature a gregarious animal: he lives in groups or communities. The most widespread community unit was, and in many places still is, the village. In primitive societies, a self-supporting community settles and sets up a village at a spot where food, water, fuel, and building materials are all at hand, though religious and behavioural factors can be important in site location. With social evolution and contact between people living in different communities, the possibility arises of one community serving others by making goods and providing services in exchange for money or money equivalents such as food: economic forces evolve which play an important, some would say over-riding role in the siting of settlements. In this opening chapter we shall explore the influence of both economic and physical factors on settlement patterns.

The Economic Influence

The process of specialization, one community or part of a community concentrating on producing goods and services for others, may trigger the formation of towns and cities which then grow of their own accord. Consequently a hierarchy of settlements develops, consisting of hamlet, village, town, and city. In the absence of physical constraints, the distribution of settlements can be explained solely in terms of economics.

In 1933, Walter Christaller proposed that, given an isotropic landscape (one which is physically uniform and has an even distribution of resources, population, wealth and the like), settlements would be evenly distributed in order to render most efficient the supplying of goods and services by people living in central settlements (central places) to people living in surrounding communities. The central places and surrounding consumer communities are linked by flows of money and goods which result from the demand of the consumers being met by the supply of the producers — butchers, bakers, and so on — in the central places.

12

SUPPLY, DEMAND AND CENTRAL PLACES

To obtain goods and services, a consumer from one of the surrounding communities must travel to a central place. This movement uses up time, energy, and money in overcoming the friction of distance. At a certain distance from the central place, the cost of travelling to the central place, on top of the cost of goods and services on arrival, is so great that consumer demand is reduced to zero; this distance is called the range of a good. Different goods have different ranges: consumers will travel far, though infrequently, to buy furniture, whereas they will travel a small distance, frequently, to buy bread. Furniture is an example of a high-order good and bread is an example of a low-order good. Any good or service also has a threshold of demand or size of market or market area which is the minimum number of consumers required to support a business (Figure 1.1). A local grocery store has a low threshold whereas a department store has a high threshold.

nb out of town s-cs

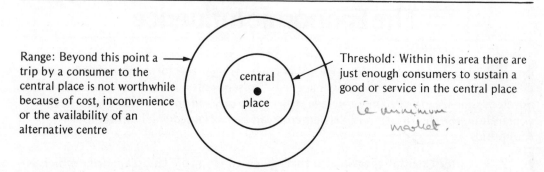

Range: Beyond this point a trip by a consumer to the central place is not worthwhile because of cost, inconvenience or the availability of an alternative centre

central place

Threshold: Within this area there are just enough consumers to sustain a good or service in the central place

le minimum market.

Figure 1.1   The threshold and range of a good or service

In a market situation of perfect competition (one in which no individual can influence the prices of goods and services), price reflects the balance between demand and supply, that is, the willingness of people to buy and sell goods at different prices. The higher the price, the less the quantity demanded; the lower the price, the less suppliers are prepared to sell. Theoretically, there is an equilibrium point at which consumers are prepared to buy the same amount as suppliers are prepared to sell; on a graph this is the point where the demand and supply curves intersect and this is therefore the market price.

1 With the aid of data in Table 1.1 construct a demand and supply schedule on graph paper (Figure 1.2) and label each curve.

2 What is the market price for this commodity? *£4.15*

3 The amount of a given commodity that any individual is prepared to buy depends not only on the market price but also on the cost of travelling to and from the

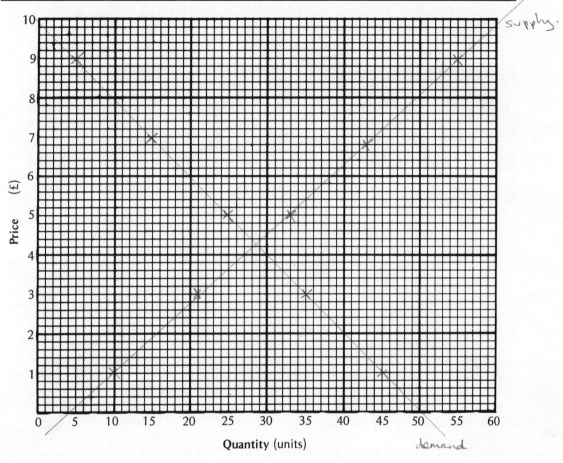

supply.

demand.

Figure 1.2

central place. Assuming that return travel costs are 50 pence per kilometre, use your graph to calculate the range, remembering that this demarcates the boundary beyond which the demand for the good is zero.   8km.

Table 1.1   Demand and supply data

Price per unit (£)	Demand per unit time	Supply per unit time
1	45	10
3	35	21
5	25	33
7	15	43
9	5	55

14

To ensure that every consumer is supplied, the market areas of competing central places overlap and a hexagonal pattern is produced (Figure 1.3).

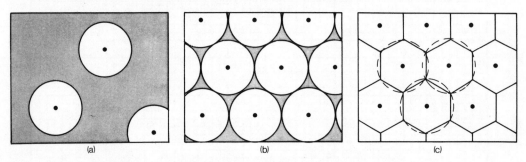

(a)   (b)   (c)

Figure 1.3   Stages in the development of hexagonal market areas
                      (a) Isolated central places in an area of low population density
                      (b) Population growth leads to increase in number of central places — market areas touch
                      (c) Further population growth means that, to pack central places even more densely, market areas have to overlap and hexagons are produced

The network of central places shown in Figure 1.4 is known rather cryptically as a K = 3 network. The K-value is determined by the number of communities of low-order places served by a central place or higher-order place. In the simplified case (Figure 1.4a) it can be seen that a first-order centre provides goods and services for one-third of the population of six consumer communities, as well as its own population; it thus serves the population equivalent of three settlements so K = 3.

The whole system of central places forms a hierarchy not only because low-order goods are provided by low-orders centres and high-order goods are provided by high-order centres, but also because the population of consumer communities served increases with increasing order of centre (Figure 1.4b).

4   To illustrate the last point, refer to Figure 1.4b and complete Table 1.2 by calculating the number of centres served by each of the orders of settlement.

Table 1.2

Order of Settlement	Number of centres served		
	All	Part	Total
1			
2			
3			
4			

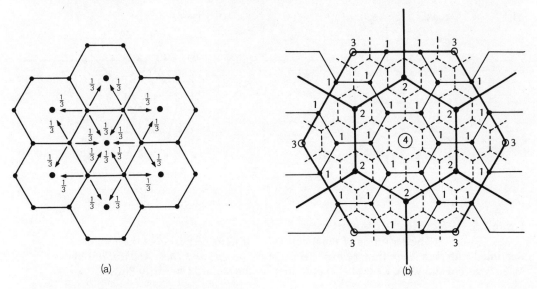

Figure 1.4   Central place networks (K = 3)
                   (a) The derivation of the 3 in K = 3
                   (b) The K = 3 hierarchy
(a) Reprinted with permission from *Human Geography: Theories and Their Applications* by
M. G. Bradford and W. A. Kent (1977) published by the Oxford University Press.

The K = 3 network is geared to efficient marketing and in its most simple form
(Figure 1.4a) allows consumers in the lower order centres to choose between three
competing central places. Other network structures are possible in which the settlement pattern is more economically efficient for travelling between places and for
administering them. The K = 4 network is the most economical arrangement for
traffic flow (Figure 1.5a): lower-order centres lie along straight lines between higher-
order centres.

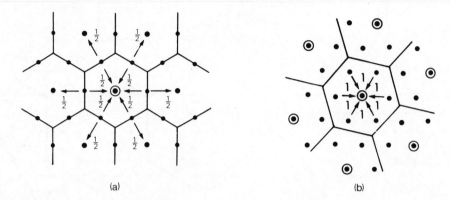

(a)   (b)

Figure 1.5   More central place networks
              (a) The derivation of the 4 in K = 4   (b) The derivation of the 7 in K = 7
Reprinted with permission from *Human Geography: Theories and Their Applications* by
M. G. Bradford and W. A. Kent (1977) published by the Oxford University Press.

Each central place in this case serves one-half the population of six lower-order
centres and its own population; thus in total it serves $(6 \times \frac{1}{2}) + 1 = 4$ places; therefore
K = 4. For the purpose of administration, a larger hexagonal unit which includes all
six lower-order centres (Figure 1.5b) is economically more efficient to run than the
K = 3 or K = 4 systems since lower-order centres would have to be jointly adminis-
tered by three or two higher-order centres respectively. In this network arrangement,
a central place serves its own population plus the entire population of six lower-order
centres; therefore K = 7.

Higher K-value systems can be derived from these basic networks. For example, the
K = 9 and K = 21 networks are based on the marketing principle (Figure 1.6).

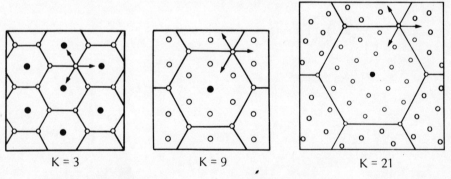

K = 3   K = 9   K = 21

Figure 1.6   Extensions of the K = 3 hierarchy (marketing principle): K = 9, K = 21

Christaller argued that, once established, the K system remained fixed. Therefore, in a system based on the marketing principle, the smallest threshold size or trading area would be K = 3, the next largest would be K = 9 and so on. Even though the high-order centres contain all the functions of the smaller centres, this produces a very marked hierarchy of functions and regular distribution of centres of the same order (Figure 1.7a and 1.7b).

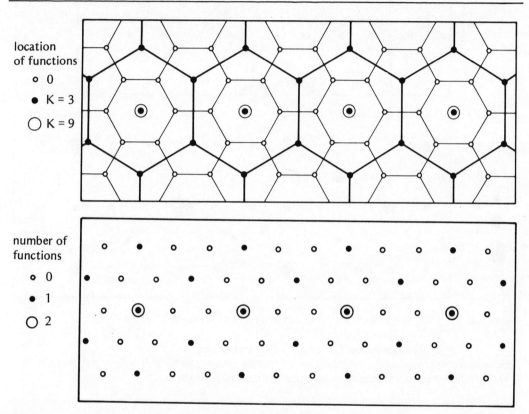

Figure 1.7   The hierarchy of central places in a fixed K network (K = 3)
(a) The K = 3 network
(b) The number of functions in each settlement

Reprinted with permission from *Techniques in Human Geography* by P. Toyne and P. Newby (1971) published by Macmillan, Basingstoke and London, figure 2.

An important modification to Christaller's scheme was made by August Lösch in 1939 who used all K networks together and varied their size. The K = 3 system was used for the commodity with the lowest threshold requirement, then the K = 4 for the next largest threshold requirement, then K = 7 and so on. This results in an irregular distribution of centres of the same order (Figure 1.8a and 1.8b) in which high-order centres do not necessarily contain all the functions of lower-order centres.

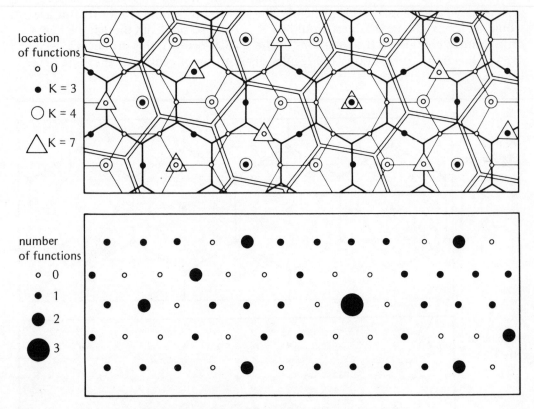

Figure 1.8   The hierarchy of central places in a relaxed K network
(a) The relaxed K
(b) The number of functions in each settlement
Reprinted with permission from *Techniques in Human Geography* by P. Toyne and P. Newby
(1971) published by Macmillan, Basingstoke and London, figure 2.

Lösch orientated all these nets about a common centre to find the position in which
the largest number of locations coincide and the distances between centres are
minimal, and so produced a Löschian landscape which shows a pattern of sectors
around a city (Figure 1.9).

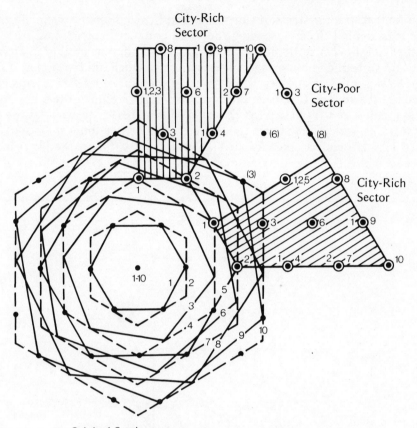

Figure 1.9 The ten smallest possible market areas
Reprinted with permission from *The Economics of Location* by A. Lösch (1954) published by
Gustav Fischer Verlag, figure 27.

THE SIZE, SPACING AND NUMBER OF SETTLEMENTS

The simplified, abstract systems of Christaller and Lösch do help to explain the economic logic behind the size, spacing and number of settlements. In Christaller's scheme, the lowest-order settlements are equidistant from one another in an isotropic landscape and surrounded by hexagonal service areas which, unlike circles, do not overlap or leave some areas unserved. For every six lowest-order settlements is a larger settlement which in turn is equidistant from settlements of the same order and so on. The smallest centres lie approximately 7 kilometres apart, whereas the centres of the next order, which serve three times the area and three times the population, are located 12.1 kilometres apart ($\sqrt{3} \times 7$). Consequently, the marketing principle, as we have seen already, gives maximum choice of central places to individual sub-centres (Figure 1.10).

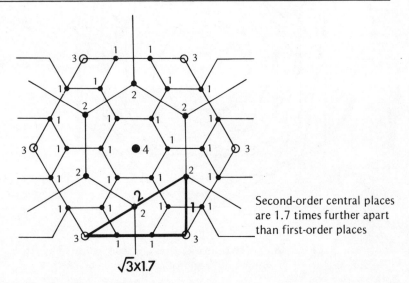

Second-order central places are 1.7 times further apart than first-order places

$\sqrt{3} \times 1.7$

Figure 1.10   Spacing in a K = 3 settlement hierarchy
Reprinted with permission from *Introducing Towns and Cities* by K. Briggs (1974) published by Hodder & Stoughton Educational, figure 11d.

To derive his hierarchy, Christaller used population data from Southern Germany. When applying his model to other areas, a lack of coincidence between census areas and the settlement limits may render such data difficult to work with. Instead, as the function of a central place is to provide central goods and services, it would seem reasonable to derive a hierarchy using an index based on the goods and services provided. Kenneth Briggs (1974) has suggested a classification of services according to their order based on their centrality index:

$$\text{Centrality Index} = \frac{100}{\text{Number of places where service is available}}$$

A high-order service will be found in just a few centres and therefore will have a high centrality value.

1 Working in pairs and using data from Table 1.5, complete the centrality index column in Table 1.5. Those of the first twelve settlements have already been worked out for you. Simply multiply the centrality index by the number of functions for each type of function and sum these results.

Example: Chillerton (reference number 48)
Centrality Index = (2.2 x 1) + (2.4 x 1) = 4.6

2 Construct a rank-size graph on semi-logarithmic section paper (Figure 1.11) to determine different orders of settlements. These are revealed by distinct breaks or kinks in the curve.

3 How well does the rank-size graph support the notion of a hierarchy of settlement in the Isle of Wight? You should be able to identify approximately six orders of settlement.

In Christaller's scheme, the number of central places which are expected in each order can be derived theoretically. For example, for a K = 3 network with a total of 81 settlements, the predicted number of central places of each order is, to the nearest whole number:

1st order = 81 − 81/3   = 54
2nd order = 81/3 − 81/9   = 18
3rd order = 81/9 − 81/27   = 6
4th order = 81/27 − 81/81   = 2
5th order = 81/81 − 81/243 = 1

4 Complete Table 1.3 for the Isle of Wight. Compare your predicted results with those of the graph and attempt to give reasons for the differences between the theoretical results and those of the graph.

Table 1.3

K = 3		K = 4		Rank size graph results
1st	55 − 55/3		55 − 55/4	
2nd	55/3 − 55/9		55/4 − 55/16	
3rd	55/9 − 55/27		55/16 − 55/64	
4th	55/27 − 55/81		55/64 − 55/256	
		55		55

The hierarchial structure of settlements exhibits an identifiable pattern in the number and type of function provided. B. J. L. Berry and W. Garrison have discovered a significant correlation between the size of settlements and the number of retailing and service outlets.

5 With reference to Table 1.5, construct a scattergraph on log-log graph paper (Figure 1.12) to show the relationship between population size and the total number of functions. Draw the line of best fit (this will follow the general trend of the points). The slope of the line shows the increase in number of functions per unit increase in population. What is the rate of increase shown on your graph?

22

6 How useful would this analysis be if one were planning the provision of services in a newly settled area?

7 So far we have examined all functions together, but we might expect a difference between functions. Again, using logarithmic section paper (Figure 1.13), select two different functions from Table 1.6 and examine their population—number of functions relationship.

8 The best-fit line can also be used to predict the minimum number of people required to support a function, that is, its minimum threshold requirement. Given the two functions you have already selected, examine your graphs and predict the population size required to support 1, 5, 10, 15 establishments (where applicable) (Table 1.4). Compare these results with the actual data in Table 1.6.

Table 1.4

Number of establishments	Establishment A		Establishment B	
	Graph	Table 1.5	Graph	Table 1.5
1				
5				
10				
15				

Table 1.5

Reference Number	Settlement	Population	Number of Functions	Centrality Index	Rank Order
1	Newport	18702	233	1772.2	1
2	Ryde	17991	236	1496.2	2
3	Shanklin	7496	169	976.2	3
4	Cowes	7212	165	952.5	4
5	Sandown	5110	127	682.6	6
6	Ventnor	5031	122	769.3	5
7	Freshwater	4070	89	549.7	7
8	E. Cowes	3986	69	344.9	8
9	Bembridge	3272	40	149.4	12
10	Totland	2950	44	213.6	9
11	Lake	2896	33	191.7	10
12	Seaview	2225	31	176.1	11
13	Carisbrook	1872	20		
14	Niton	1742	12		
15	Northwood	1512	10		
16	Wootton Bridge	1423	21		

Reference Number	Settlement	Population	Number of Functions	Centrality Index	Rank Order
17	Brading	1323	20		
18	Binstead	1305	9		
19	St. Helens	1201	13		
20	Wroxhall	1189	7		
21	Godshill	1067	16		
22	Yarmouth	853	24		
23	Brighstone	810	9		
24	Nettlestone	809	7		
25	Parkhurst	611	9		
26	Arreton	609	7		
27	Whitwell	581	5		
28	Shalfleet	551	4		
29	Shorwell	538	4		
30	Newchurch	534	5		
31	Calbourne	489	5		
32	Havenstreet	445	4		
33	Newbridge	394	3		
34	Porchfield	350	4		
35	St. Lawrence	279	3		
36	Gurnard	268	2		
37	Whippingham	250	3		
38	Apse Heath	238	3		
39	Chale	235	4		
40	Wellow	225	4		
41	Brook	218	2		
42	Winford	209	2		
43	Yaverland	180	2		
44	Chale Green	167	4		
45	Whitely Bank	153	2		
46	Rookley	150	2		
47	Blackgang	134	2		
48	Chillerton	103	2		
49	Mottistone	80	1		
50	Ningwood	78	1		
51	Thorley Street	75	0		
52	Blackwater	69	0		
53	Gatcombe	68	0		
54	Moortown	62	0		
55	Newtown	60	0		

Table 1.6

Function or service	Centrality value	Newport	Ryde	Shanklin	Cowes	Sandown	Ventnor	Freshwater	E. Cowes	Bembridge	Totland	Lake	Seaview	Carisbrook	Niton	Northwood	Wootton Br.	Brading	Rinstead	St. Helena	Wroxall	Godshill	Yarmouth	Brighstone
1 General stores	2.2	18	19	9	12	1	6	5	3	3	4	1	3	3	2	5	4	4	2	2	2	3	—	2
2 Newsagents	4.4	9	7	4	3	3	1	4	8	1	2	2	1	1	2	2	3	2	1	2	2	2	4	1
3 Public house	2.4	17	20	13	15	13	13	4	3	7	9	3	2	1	1	1	1	2	1	1	1	1	1	1
4 Post Office	2.4	8	7	3	4	3	1	3	2	1	1	1	1	1	—	1	—	—	—	—	—	—	1	—
5 Baker	4.2	4	3	3	3	1	1	1	3	—	1	1	1	—	1	—	—	—	—	—	—	—	1	—
6 Butcher	6.3	7	7	7	10	3	4	1	1	1	—	1	1	1	—	1	1	—	—	—	—	—	2	—
7 Garage	1.5	6	7	4	5	8	5	4	2	5	2	2	1	4	2	1	4	1	1	1	1	—	2	1
8 Hardware	6.3	12	12	3	6	3	3	7	6	—	1	3	1	2	—	—	1	1	1	—	—	—	1	—
9 Grocer	6.3	11	5	2	3	3	3	1	—	2	2	1	1	1	—	—	—	—	—	1	—	—	1	—
10 Antiques	2.6	6	3	3	8	2	9	1	1	1	2	1	—	2	—	—	—	—	—	—	1	—	1	—
11 Bank	6.3	4	5	4	2	3	5	5	4	3	1	1	3	—	1	—	—	1	—	1	—	—	1	—
12 Book shop	16.7	3	3	2	2	2	1	—	—	—	—	—	—	—	—	—	—	—	—	—	1	—	—	—
13 Builders	25.0	2	—	1	—	—	1	—	—	—	—	—	—	—	—	—	—	—	—	—	—	—	—	—
14 Cafe	0.8	5	14	21	14	20	10	5	7	2	6	—	3	—	—	—	—	1	1	—	—	5	4	2
15 Camera	33.3	2	2	—	—	—	1	—	—	—	—	—	—	—	—	—	—	—	—	—	—	—	—	—
16 Chemist	7.1	6	6	6	3	5	3	3	2	2	1	1	1	1	1	1	—	1	—	—	—	—	—	—
17 Clothes	5.6	10	13	13	16	6	1	6	4	1	3	3	3	1	—	1	1	—	—	2	—	—	—	—
18 Dry cleaner	11.1	3	4	2	3	—	1	1	1	—	—	1	—	—	—	—	—	—	—	—	—	—	—	—
19 TV—Electrical	9.1	15	8	5	4	2	1	2	1	—	—	2	—	2	—	—	—	—	—	—	—	—	—	—
20 Estate agent	8.3	7	8	4	4	5	5	2	2	2	2	—	2	—	—	—	2	—	—	—	—	—	—	—
21 Florist	16.7	4	3	2	2	—	2	—	1	—	—	—	—	—	—	—	—	—	—	—	—	—	—	—
22 Fishmonger	9.1	3	2	1	—	—	2	2	—	—	—	—	—	—	—	—	—	1	—	—	—	—	1	—
23 Furniture	12.5	9	7	3	2	2	2	3	1	—	—	—	—	—	—	—	—	—	—	—	—	—	—	—
24 Gift	6.7	8	15	20	8	12	12	7	4	—	7	—	1	—	—	—	—	5	—	—	—	4	—	2
25 Greengrocer	7.7	3	3	3	1	1	2	4	2	—	2	1	1	—	—	—	—	1	1	2	—	—	1	—
26 Hairstylist	2.0	7	11	3	7	4	4	3	3	2	—	1	—	—	—	—	—	1	1	2	—	—	1	—
27 Jeweller	11.1	3	—	5	3	5	2	2	—	—	—	2	1	—	—	1	—	—	—	1	—	—	—	—
28 Launderette	14.3	3	4	2	1	2	1	1	1	—	—	—	1	—	—	—	—	—	—	—	—	—	—	—
29 Leather	14.3	1	1	1	3	1	1	1	1	—	—	—	—	—	—	—	—	—	—	—	—	—	—	—
30 Optician	16.7	3	2	1	—	1	1	1	1	—	—	—	—	—	—	—	—	—	—	—	—	—	—	—
31 Photo studio	16.7	2	1	—	3	—	2	2	—	—	—	—	—	—	—	—	—	—	—	—	—	—	1	—
32 Records	25.0	2	1	—	—	—	—	—	—	—	1	—	—	—	—	—	—	—	—	—	—	—	—	—
33 Restaurant	8.3	4	3	3	1	3	2	1	—	1	—	—	—	—	—	—	1	—	—	—	—	—	1	1
34 Shoes	11.1	4	5	3	—	2	2	—	2	—	—	2	1	—	—	1	—	—	—	—	—	—	—	—
35 Sports Equipment	2.0	3	2	1	—	—	1	1	—	—	—	—	—	—	—	—	—	—	—	—	—	—	—	—
36 Stationers	11.1	2	1	1	—	2	2	—	—	1	—	1	—	—	—	1	—	—	—	—	—	—	1	—
37 Supermarket	9.1	5	3	2	4	1	4	4	1	1	—	2	1	—	—	—	—	—	—	—	—	—	—	—
38 Toys	14.3	5	4	2	—	2	—	1	—	—	—	—	1	—	—	—	—	—	—	—	—	—	—	—
39 Travel agent	8.3	3	2	1	4	—	—	1	—	—	1	—	—	1	—	—	—	—	—	—	—	—	—	—
40 Solicitors	2.8	9	7	3	6	4	2	2	—	1	1	—	1	—	—	—	—	—	—	—	—	—	—	—
41 Undertakers	20.0	2	—	—	2	—	1	—	—	1	2	—	1	—	—	—	—	—	—	—	—	1	—	—
42 Wine store	6.3	2	4	1	1	1	2	—	—	1	2	—	1	—	—	—	—	—	—	—	—	—	—	—
43 Cinema	16.7	2	2	1	—	1	—	—	—	—	—	—	—	—	—	—	—	—	—	—	—	—	—	—
Total Number of Functions		233	236	169	165	127	122	89	69	40	44	33	31	20	12	10	21	20	9	13	7	16	24	9

Nettlestone	Parkhurst	Arreton	Whitwell	Shalfleet	Shorewell	Newchurch	Calbourne	Havenstreet	Newbridge	Porchfield	St. Lawrence	Gurnard	Whippingham	Apse Heath	Chale	Wellow	Brook	Winford	Yaverland	Chale Green	Whitley Bank	Rookley	Blackgang	Chillerton	Mottistone	Ningwood	Thorley St.	Blackwater	Gatcombe	Moortown	Newtown
2	1	1	2	1	1	1	1	1	1	1	1	–	1	1	1	1	–	1	1	1	2	1	–	1	1	–	–	–	–	–	–
–	1	1	–	–	–	–	–	–	–	–	–	–	–	1	–	–	–	–	1	–	–	–	1	–	–	–	–	–	–	–	–
2	1	1	1	1	1	1	1	1	1	1	1	2	1	1	–	1	1	–	–	–	1	–	–	–	–	–	1	–	–	–	–
1	1	1	1	–	1	1	1	1	1	1	1	1	–	1	1	1	1	1	–	–	1	–	1	–	1	–	–	–	–	–	–
–	–	–	–	–	–	–	–	–	–	–	–	–	–	–	–	–	–	–	–	–	–	–	–	–	–	–	–	–	–	–	–
–	–	–	–	–	–	–	–	–	–	–	–	–	–	–	–	–	–	–	–	–	–	–	–	–	–	–	–	–	–	–	–
1	1	1	1	1	1	1	1	1	1	–	1	–	–	–	1	–	–	–	–	1	–	–	–	–	–	–	–	–	–	–	–
–	1	–	–	–	–	–	–	–	–	–	–	–	–	–	–	–	–	–	–	–	–	–	–	–	–	–	–	–	–	–	–
–	–	–	–	–	–	1	–	–	–	–	–	–	–	–	–	–	–	1	–	–	–	–	–	–	–	–	–	–	–	–	–
–	–	–	–	–	–	–	–	–	–	–	–	–	–	–	–	–	–	–	–	–	–	–	–	–	–	–	–	–	–	–	–
–	–	–	–	–	–	–	–	–	–	–	–	–	–	–	–	–	–	–	–	–	–	–	–	–	–	–	–	–	–	–	–
–	1	–	–	–	–	1	–	–	–	–	–	–	–	–	–	1	1	–	–	–	–	–	–	1	–	–	–	–	–	–	–
–	–	–	–	–	–	–	–	–	–	–	–	–	–	–	–	–	–	–	–	–	–	–	–	–	–	–	–	–	–	–	–
1	–	–	–	1	–	–	–	–	–	–	–	–	–	–	–	–	–	–	–	–	–	–	–	–	–	–	–	–	–	–	–
–	1	–	–	–	–	–	–	–	–	–	–	–	–	–	–	–	–	–	–	–	–	–	–	–	–	–	–	–	–	–	–
–	1	–	–	–	–	–	–	–	–	–	–	–	–	–	–	–	–	–	–	–	–	–	–	–	–	–	–	–	–	–	–
–	–	–	–	–	–	–	–	–	–	–	–	–	–	–	–	–	–	–	–	–	–	–	–	–	–	–	–	–	–	–	–
–	–	–	–	–	–	–	–	–	–	–	–	–	–	–	–	–	–	–	–	–	–	–	–	–	–	–	–	–	–	–	–
–	–	1	–	–	–	–	–	–	–	–	–	–	–	–	–	–	–	–	–	–	–	–	1	–	–	–	–	–	–	–	–
–	1	–	–	–	–	–	–	–	–	–	–	–	–	–	–	–	–	–	–	–	–	–	–	–	–	–	–	–	–	–	–
–	–	–	–	–	–	–	–	–	–	–	–	–	–	–	1	–	–	–	–	–	–	–	–	–	–	–	–	–	–	–	–
–	–	–	–	–	–	–	–	–	–	–	–	–	–	–	–	–	–	–	–	–	–	–	–	–	–	–	–	–	–	–	–
–	–	–	–	–	–	–	–	–	–	–	–	–	–	–	–	–	–	–	–	–	–	–	–	–	–	–	–	–	–	–	–
–	–	–	–	–	–	–	–	–	–	–	–	–	–	–	–	–	–	–	–	–	–	–	–	–	–	–	–	–	–	–	–
–	–	1	–	–	–	1	–	–	–	–	–	–	–	–	–	–	–	–	–	1	–	–	–	–	–	–	–	–	–	–	–
–	–	–	–	–	–	–	–	–	–	–	–	–	–	–	–	–	–	–	–	–	–	–	–	–	–	–	–	–	–	–	–
–	–	–	–	–	–	–	–	–	–	–	–	–	–	–	–	–	–	–	–	–	–	–	–	–	–	–	–	–	–	–	–
–	–	–	–	–	–	–	–	–	–	–	–	–	–	–	–	–	–	–	–	–	–	–	–	–	–	–	–	–	–	–	–
–	–	–	–	–	–	–	–	–	–	–	–	–	–	–	–	–	–	–	–	–	–	–	–	–	–	–	–	–	–	–	–
–	–	–	–	–	–	–	–	–	–	–	–	–	–	–	–	–	–	–	–	–	–	–	–	–	–	–	–	–	–	–	–
–	–	–	–	–	–	–	–	–	–	–	–	–	–	–	–	–	–	–	–	–	–	–	–	–	–	–	–	–	–	–	–
7	9	7	5	4	4	5	5	4	3	4	3	2	3	3	4	4	2	2	2	4	2	2	2	2	1	1	0	0	0	0	0

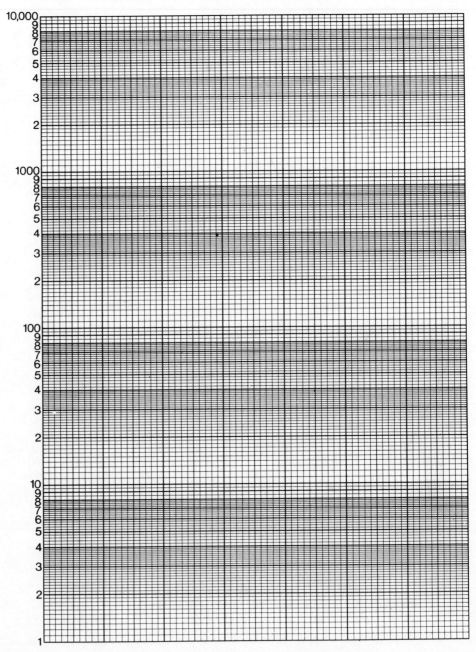

Figure 1.11

Note: This figure uses a logarithmic scale on the *y* axis. This differs from the graph paper with which you are most familiar since this scale increases at a geometric rate (whereas the *x* axis increases at an arithmetic rate), that is, it increases at a constant rate of ten and is divided into cycles, the end of each cycle being ten times greater than the end of the previous one.
Example: 0.01; 0.1; 1.0; 10.0; and so forth.

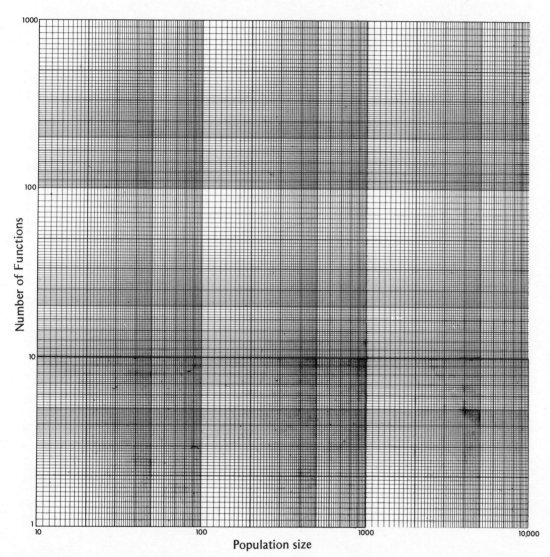

Figure 1.12

Note: This figure uses the logarithmic scale on both axes.

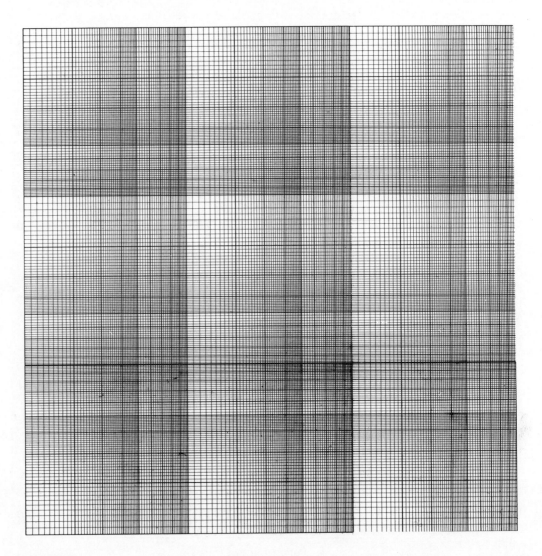

Figure 1.13

THE SPACING OF SETTLEMENTS

Given a hierarchy of settlement, it is reasonable to suppose that the average spacing between centres of each order will become greater with increasing order of settlement. According to Christaller's model, in a K = 3 network, the spacing of each higher order of centre should on average be 1.7 times as great as that of the order below (see p. 20).

By convention, predicted spacing is based on that of the lowest order of centre. Thus, with a 1st-order spacing of 5 kilometres, it follows that

2nd-order spacing = $5 \times \sqrt{3}$ = 8.7 km
3rd-order spacing = $8.6 \times \sqrt{3}$ = 15 km
4th-order spacing = $15 \times \sqrt{3}$ = 26 km
5th-order spacing = $26 \times \sqrt{3}$ = 45 km

1 For each order of settlement that you have already identified in the Isle of Wight, measure the distance between nearest neighbours (see Figure 1.15) within that category and calculate the average spacing.

2 Calculate the theoretical spacing as shown above and compare your results in Table 1.7.

Table 1.7

Order of settlement	Spacing (km)	
	Isle of Wight	Theory
1st		
2nd		
3rd		
4th		
5th		

Suggest reasons for similarities and differences.

As we saw earlier (p. 16), Christaller also incorporated an administrative principle into his model where K = 7. In this scheme, all six lower-order places owe allegiance to just one higher-order centre instead of sharing as in the previous examples. Each hexagon of any given order is naturally in contact with six other hexagons of equal importance. Thus if we examine the areas served by central places they should be in contact with six similar areas. Peter Haggett has shown that in Brazil, municipios (the basic Brazilian administrative area of which there are approximately 2800), according to sample survey, are in contact with 5.71 other municipios. We shall explore this further in the next chapter.

THE NEAREST-NEIGHBOUR STATISTIC AS A GUIDE TO SPACING

By using what is called nearest-neighbour analysis, the hexagonal arrangement of settlements can be tested more rigorously. The technique of nearest-neighbour analysis produces a value, R_n, which ranges from 0 to 2.15 (Figure 1.14) and describes the dispersion of settlements. In our analysis we should expect uniform dispersion.

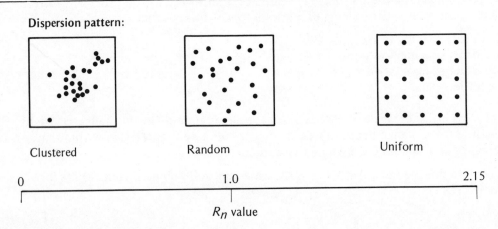

Figure 1.14   The nearest neighbour, R_n, scale

1   Refer to Figure 1.15 and for each settlement measure the distance from its nearest neighbour irrespective of order. Note that it is possible to measure back where reflexive pairs occur (Figure 1.16).

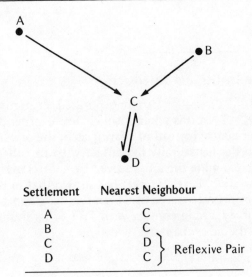

Settlement	Nearest Neighbour
A	C
B	C
C	D
D	C

} Reflexive Pair

Figure 1.16   Measuring nearest neighbours

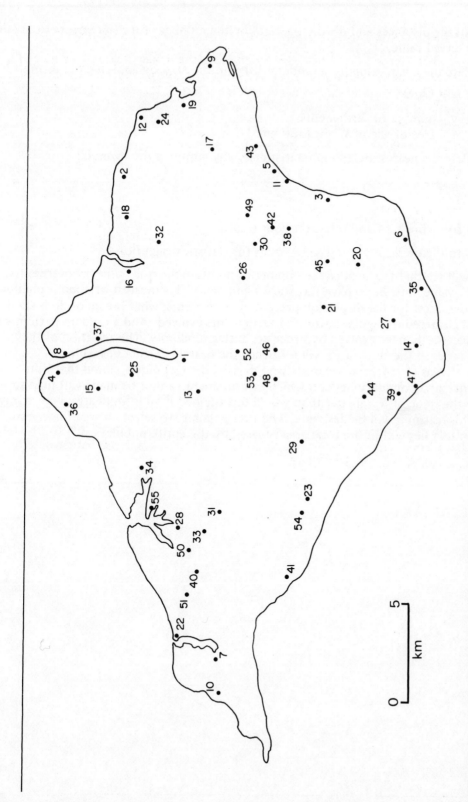

Figure 1.15   Distribution of settlements on the Isle of Wight

2   Sum all the distances and divide this total by the number of measurements to obtain the observed value, $\bar{D}_{obs}$.

3   Calculate the mean value for a random distribution, $\bar{D}_{ran}$ by applying the formula

$$\bar{D}_{ran} = 0.5\sqrt{A/N}$$

where N = number of settlements
A = area of Isle of Wight (380 km^2).

4   Calculate the nearest neighbour statistic, R_n, by applying the formula:

$$R_n = \frac{\bar{D}_{obs}}{\bar{D}_{ran}}$$

5   Refer to Figure 1.14 and interpret your result.

6   Refer to Figure 1.17 and test your result for statistical significance.

You will see that the smaller the number of points in the study area the more extreme the R_n value must be to have statistical significance. The nearest-neighbour method provides a test for the dispersion pattern of settlements; what it cannot do is show how a clustered or regular pattern of settlements evolved. And a random pattern will not necessarily have evolved by processes acting randomly. The selection of the boundaries of the study area will influence the nearest-neighbour results: P. Clark and F. Evans (1954), the botanists who devised the technique, stated that "the presence of a boundary beyond which measurement cannot be made will tend to make the value of R_n greater than would be obtained if an infinite area were involved. For this reason it will be desirable, wherever possible, to select an area for investigation that lies within the total area covered by the entire population."

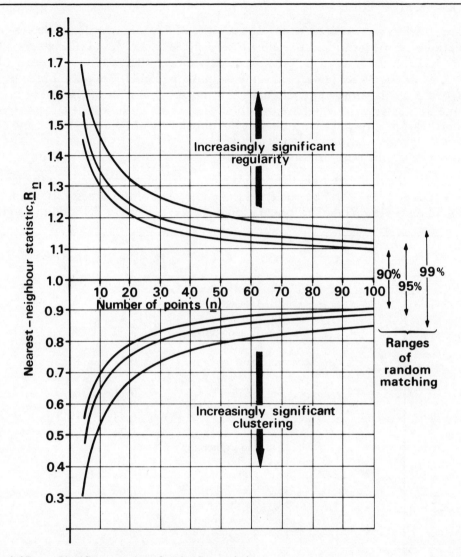

Figure 1.17 Significance values for the R_n statistic
Reprinted with permission from *The elimination of under-estimation in nearest-neighbour analysis* by D. A. Pinder (1978), Department of Geography, University of Southampton, *Discussion Paper No. 1*, figure 7.

THE RANK-SIZE RULE

Cities themselves vary in size. Arranging the cities of a region in order of decreasing population size and plotting this against rank order produces a rank-size graph (Figure 1.18). As early as 1913, Felix Auerbach noticed that, arranged in order of size, the population sizes of cities in some regions are related. Basically, the population of the second-largest city will have half the population of the largest or primate city; the third-largest city will have one-third of the population of the largest city; the fourth-largest city will have a quarter the population of the largest city; and so forth. This inverse relationship between city population and its rank order in a set of cities is called the rank-size rule; it was given a firm footing by G. K. Zipf in 1949.

According to the rank-size rule, the population of the third-largest city in the United States — Los Angeles — will be one-third the population of the largest city in the United States — New York. The population of New York city is 7.78 million and so

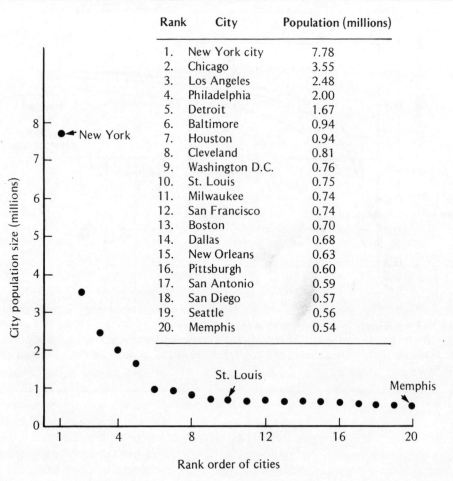

Rank	City	Population (millions)
1.	New York city	7.78
2.	Chicago	3.55
3.	Los Angeles	2.48
4.	Philadelphia	2.00
5.	Detroit	1.67
6.	Baltimore	0.94
7.	Houston	0.94
8.	Cleveland	0.81
9.	Washington D.C.	0.76
10.	St. Louis	0.75
11.	Milwaukee	0.74
12.	San Francisco	0.74
13.	Boston	0.70
14.	Dallas	0.68
15.	New Orleans	0.63
16.	Pittsburgh	0.60
17.	San Antonio	0.59
18.	San Diego	0.57
19.	Seattle	0.56
20.	Memphis	0.54

Figure 1.18   Rank-size graph for twenty largest United States cities

we should expect the population of Los Angeles to be 7.78/3 = 2.59 million; in fact the population of Los Angeles is 2.48 million, a good confirmation of the rank-size rule, for this case at least.

To compare the observed rank-size relation of a set of cities with the theoretical rank-size relation between them, it is helpful to plot both distributions on sectional paper which has logarithmic divisions along both axes. Figure 1.19a shows the rank-size relation predicted by theory plotted on arithmetic sectional paper — the result is a J-shaped curve. The same data plotted on double-log paper produces a straight line (Figure 1.19b). Observed rank-size data similarly plotted will usually produce a straight line, possibly with a different slope, and can be readily compared with the predicted case.

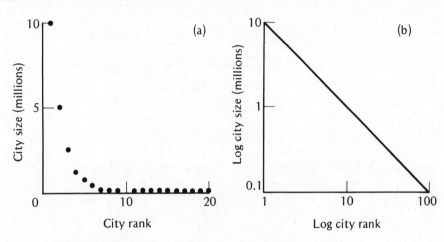

Figure 1.19 Rank-size curves
(a) J-shaped curve on arithmetic sectional paper
(b) Straight-line on log-log paper

1 Plot the rank-size data for North American cities (Figure 1.18) on double-log paper (Figure 1.20). Comment on the result and, before reading further, endeavour to think of reasons why cities should conform to this rule.

Cities in all regions do not conform to the straight-line relationship between size and rank. Australia, for instance, is dominated by a few large cities of roughly equal size and these are followed by a string of smaller cities which do conform to the rank-size rule. The result, shown in Figure 1.21, is two distinct sections to the rank-size line: an upper, rather flat section; and a lower, straight section which lies more or less parallel to the usual rank-size slope. This pattern is called binary, trinary, quaternary, and so on, depending on the number of cities in the upper section. The general term for this pattern is polynary. In Russia one large city, Moscow, dominates all others. The rank-size graph falls away far more rapidly than the predicted case. The result is a primate pattern (Figure 1.21).

36

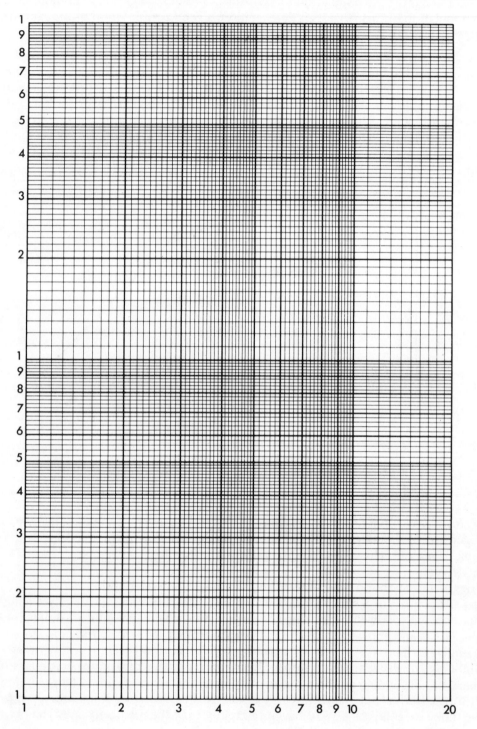

Figure 1.20

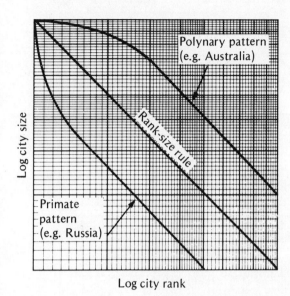

Log city size

Log city rank

Figure 1.21   Rank-size distributions

Table 1.8   (population in thousands)

France (1975)		Netherlands (1974)	
Paris	9 108	Rotterdam . . .	1 040
Lyon	1 167	Amsterdam . . .	1 002
Marseille . . .	1 004	s'Gravenhage . .	685
Lille	923	Utrecht	463
Bordeaux . . .	589	Eindhoven . . .	350
Toulouse . . .	495	Arnhem	277
Nantes . . .	433	Heerlen-Kerkrade .	265
Nice	433	Enschede-Hengelo	238
Rouen	390	Haarlem . . .	235
Grenoble . . .	389	Tilburg	211
Toulon	378	Nijmegen . . .	210
Strasbourg . . .	356	Groningen . . .	203
St. Etienne . . .	335	Dordrecht . . .	181
Lens	313	Gelcen-Sittard . .	176
Nancy	279	's-Hertogenbosch .	171
Le Havre . . .	266	Leiden	165
Grasse-Cannes . .	254	Breda	151
Tours	246	Maastricht . . .	146
Clermont-Ferrand	225	Zaandam . . .	138
Valenciennes . .	224	Velsen-Beverwijk .	137

2 Refer to Table 1.8 which lists in rank order the population of the twenty largest cities in France and the Netherlands.

  a Construct the theoretical rank-size line (Figure 1.22b) based on the population of the largest city as shown in Figure 1.22a.

  b Construct rank-size graphs (Figures 1.22a and 1.22b) for the two countries and attempt to explain the differences between them.

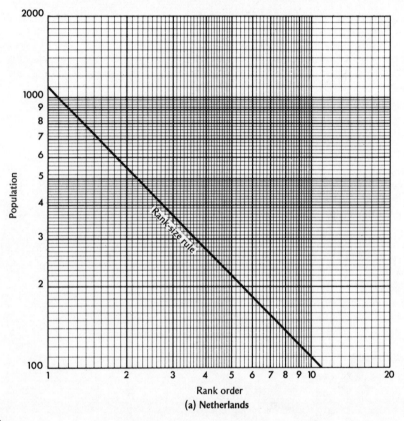

Figure 1.22a

Vapnarsky (1969) argued that, because of increasing interregional linkages within the national economy and increasing dependence on international trade as time progresses, the settlements in a country shift from a primate to a rank-size distribution. If the rank-size distribution is the one to which most countries will eventually conform, then a high degree of primacy would be expected in the less-developed countries where urbanization is a relatively recent phenomenon. A measure of the degree of primacy may be obtained by computing an index where

$$\text{Index of Primacy} = \frac{\text{1st-ranked city's population}}{\text{2nd-ranked city's population}}$$

The higher the index the higher the degree of primacy.

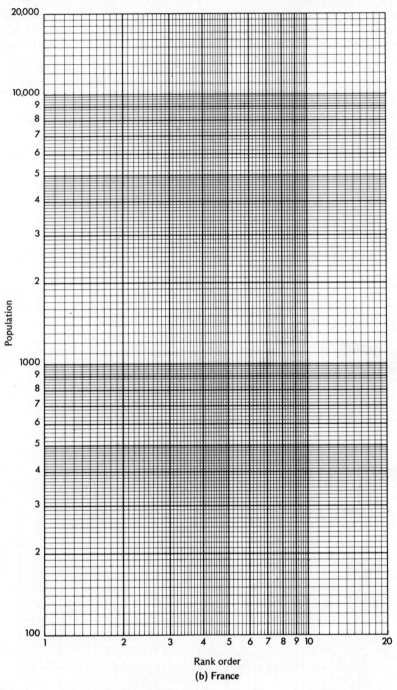

Population

Rank order
(b) France

Figure 1.22b

3   Table 1.9 lists the population of the 1st and 2nd largest cities together with rates of urbanization for seven African countries. Calculate the indices of primacy and test the hypothesis that there is a relationship between primacy and the rate of urbanization by constructing a scattergraph and drawing the line of best fit (Figure 1.23).

Table 1.9

Country	Rate of urbanization (% per year)	City population (000's)		Index of primacy
Algeria	0.78	Alger	943	
		Oran	327	
Egypt	1.33	Cairo	5715	
		Alexandria	2260	
Ghana	0.52	Accra	615	
		Kumasi	281	
Malawi	0.56	Blantyre	109	
		Zomba	19	
Morocco	1.72	Casablanca	1753	
		Rabat	596	
Rhodesia	0.32	Salisbury	390	
		Bulawayo	270	
S. Africa	0.08	Johannesburg	1152	
		Cape Town	807	

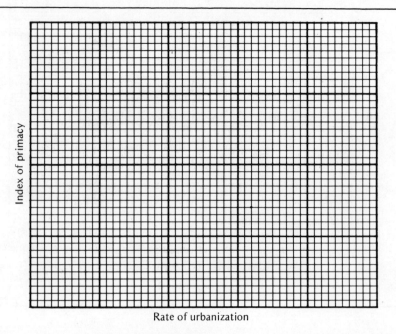

Rate of urbanization

Figure 1.23

Table 1.10

Country	Area (km^2)	Year	Index of primacy	Year	City population (000's)		Index of primacy
Argentina	2767	1950	7.9	1970	Buenos Aires	8353	
					Rosario	811	
Brazil	8512	1940	1.3	1970	São Paulo	5979	
					Rio de Janiero	4316	
Chile	752	1940	4.5	1960	Santiago	3069	
					Valparaiso	253	
Colombia	1139	1938	2.2	1973	Bogotá	2978	
					Medellín	1270	
Ecuador	284	1950	1.2	1974	Guayaquil	814	
					Quito	597	
El Salvador	21	1950	3.1	1971	San Salvador	380	
					Santa Ana	172	
Mexico	1973	1940	7.4	1970	Mexico City	10767	
					Guadalajara	1478	
Nicaragua	130	1950	3.5	1963	Managua	235	
					León	45	
Peru	1285	1940	8.0	1972	Lima	3158	
					Arequipa	305	
Venezuela	912	1941	2.9	1971	Caracas	2175	
					Maracaibo	652	

4 From what Vapnarsky said the degree of primacy declines with time. Table 1.10 provides data on selected Latin American countries; using these data calculate indices of primacy for the most recent years and complete the table. Has the degree of primacy in Latin America generally increased or decreased? From what you understand of the economic and urban development of South America, can you suggest reasons for this trend? (You may need to consult a modern systematic text on South America to answer this question.)

5 In Berry's study of 38 countries, 15 of them had primate distributions which were apparently the product of urbanization processes in countries that were small in size, experienced a recent history of urbanization, and possessed a simple economic structure. It would therefore seem reasonable to suggest countries large in terms of area and population will consist of several regions each centred on a regional capital city. In contrast, smaller countries will tend to possess a primate city structure, that is, one large city dominating a national hinterland.

With the aid of data from Table 1.10 test the hypothesis that smaller Latin American countries tend to exhibit primate city structures by calculating a coefficient of rank correlation.

A rank correlation coefficient is a value lying between −1.0 and +1.0 which indicates the strength and direction of a relationship between two variables. Positive values show that the two variables are directly related — one increases as the other increases (Figure 1.24a). A value of 0.0 or thereabouts indicates a lack of correlation between the two variables (Figure 1.24b). Negative values show an inverse relation between the two variables — one decreases as the other increases (Figure 1.24c).

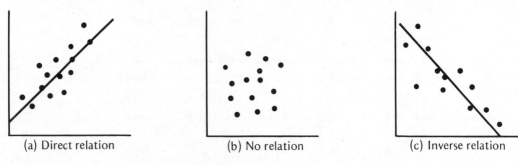

(a) Direct relation   (b) No relation   (c) Inverse relation

Figure 1.24   Direct and inverse relationships

To obtain the rank correlation coefficient, r, devised by C. Spearman, the procedure is as follows. Assume we wish to test the hypothesis that the population of the largest city in a country is related to the size of the country. Extracting the relevant data for the sample of ten Latin American countries from Table 1.10, step one is to put the observations on each of the two variables in rank order; this is done in Table 1.11, columns 3 and 5. Step two is to find the difference in ranks, d. This means subtracting values in column 5 from the values in column 3 of the table and putting the results in column 6. Step three is to square the differences and put the results in column 7. Because the square of a negative number is a positive number, the squaring gets rid of negative differences. Step four is to add up the d^2 values, to find the total of column 7. In statistical shorthand, this summing operation is written as Σd^2 where the Greek Σ means "sum of". In the example, d^2 is 16. Step five is to substitute d^2 and n, the number of pairs of observations or sample size — in our case 10, in Spearman's formula

$$r = 1 - \frac{6\Sigma d^2}{(n^3 - n)}$$

In the example this yields

$$r = 1 - \frac{16}{(10^3 - 10)}$$

$$= 1 - (16/990)$$

$$= 1 - 0.0162$$

$$= 0.984$$

The rank correlation coefficient is +0.984 which suggests a very strong, direct relationship between the population of the largest city in a country and the size of the country. The statistical significance of the result may be assessed using Figure 1.25. With a sample size of 10 and a rank correlation coefficient of 0.984, we find the relationship is significant at what is called the 0.1 percent confidence level. This means that only in one case in every thousand will the relationship have arisen by chance; in 999 cases out of a thousand there appears to be a causal connexion between the two variables. What exactly this connexion is has then to be reasoned out.

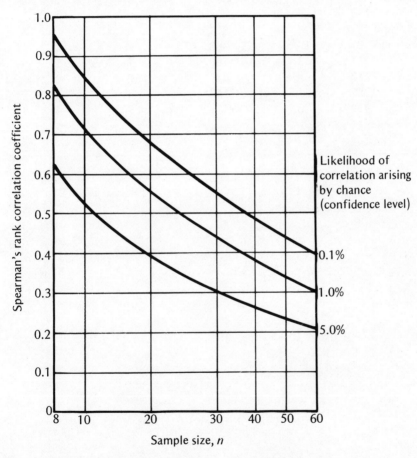

Figure 1.25   Significance values for Spearman's rank correlation coefficient
Based, with permission, on Table 6.4 from *Statistics Tables* by H. R. Neave (1978) published by George Allen & Unwin.

Table 1.11   Example data for calculating Spearman's r

1 Country	2 Area (km^2)	3 Rank area	4 Population of largest city (000's)	5 Rank population	6 d	7 d^2
Argentina	2767	2	8353	2	0	0
Brazil	8512	1	5979	3	−2	4
Chile	752	7	3069	5	2	4
Colombia	1139	5	2978	6	−1	1
Ecuador	284	8	814	8	0	0
El Salvador	21	10	380	9	1	1
Mexico	1973	3	10,767	1	2	4
Nicaragua	130	9	235	10	−1	1
Peru	1285	4	3158	4	0	0
Venezuela	912	6	2175	7	−1	1

$$\Sigma d^2 = 16$$

A CASE STUDY: THE ZUIDER ZEE PROJECT

From what we understand of the intricate web of settlement structure, we should be able to plan effectively the urban development of a rural area. The Dutch provide us with a unique example of such a planning problem in the Zuider Zee Project.

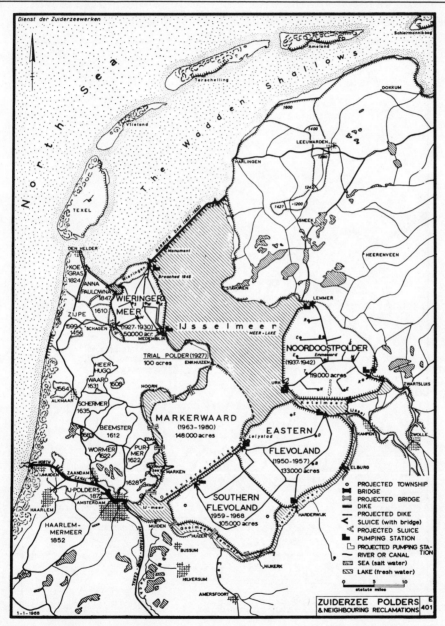

Figure 1.26   Zuider Zee Polders
Reprinted with permission from *The Zuider Zee Works* (1967) published by the Netherlands Government Information Service, figure 10.

The Wieringermeer polder (20,000 hectares) was completed in 1930. The problem then was how best to lay out settlements and lines of communications. The pattern of roads and canals would have to be largely determined by the division of agricultural land into standard 20-hectare parcels. Therefore the choice of settlement location would be fairly arbitrary, two of the villages for example being built at important road intersections. The settlement pattern which emerged proved not to be ideal: the three original villages being clustered in the middle of the polder

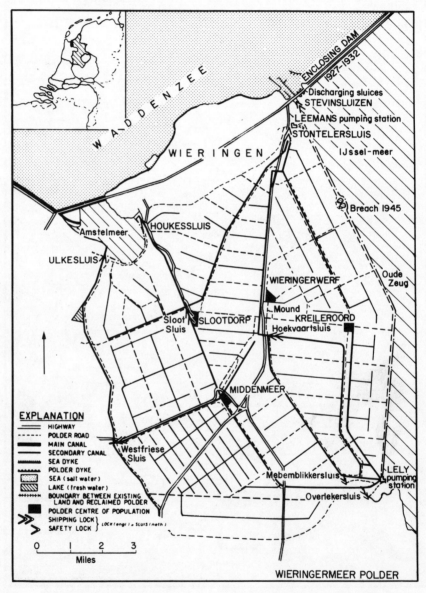

Figure 1.27 Wieringermeer Polder
Reprinted with permission from *The Zuider Zee Guide* (1975) published by the Information and Documentation Centre for the Geography of the Netherlands, figure 21.

(Figure 1.27) had overlapping service areas and people living well away from the villages were inconvenienced by long journeys. The population growth on the polder was less than the expected level and this made the disadvantages even worse: the planned population of 15,000 settlers was not attained, the area stagnating at a population of 8000 for many years, with small villages incapable of providing a satisfactory level of service. The polder is now gaining in population (12,000 in 1976) largely as a result of people from the Amsterdam agglomeration coming to live in the area and commuting to Amsterdam, 60 kilometres to the south.

Figure 1.28   Wieringerwerf surrounded by the standard 20-hectare parcels of land
Reprinted with permission from *The Zuider Zee Guide* (1975) published by the Information and Documentation Centre for the Netherlands. Photography Copyright Luchfoto Bart Hoemester, photo 10.

The polder also lacked a definite centre, a problem which became acute in 1935 when the area was about to become municipally independent. Wieringerwerf was selected as the "capital", long before building work on it was finished. Slootdorp and Middenmeer had been built two years previously and businessmen seeing a great future in the polder and not wanting to wait until Wieringerwerf was ready settled in and developed Middenmeer. As a result Wieringerwerf became the administrative centre and Middenmeer the commercial centre. Today, the lack of a proper single centre is still a problem but the development of light industry in Wieringerwerf may alter the situation. However, in an attempt to spread out the population, a fourth village has been built at Kreileroord.

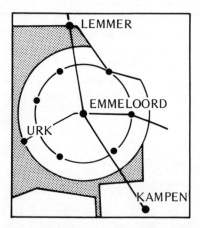

Figure 1.29   Model of the settlement pattern in the North East Polder
Reprinted with permission from *The Zuider Zee Guide* (1975) published by the Information and
Documentation Centre for the Geography of the Netherlands, figure 23.

1   Suggest ways in which central-place theory could have helped the Dutch planners to
avoid making their mistakes.

In the planning of the North East Polder (48,000 hectares) an attempt was made to
avoid repeating the mistakes made in the Wieringermeer. A systematic plan was drawn
up in which the central settlement was located at the intersection of a skeletal road
network with smaller service centres surrounding it (Figure 1.29).

But how many villages would have to be built? Initially, five were considered sufficient since former coastal villages on the mainland could serve part of the new
polderland. Doubts were raised as to whether the old villages would be suitable and
the plan changed to include six new villages. In 1948 a survey was conducted on the
old mainland to determine how far people were prepared to travel from the outlying
districts to the villages. The results suggested that 5 kilometres was the maximum
distance and, together with the fear that farmers might have difficulty attracting
labour if the journey to work were too great, the plan was changed yet again. In the
end, a decision was made to build ten villages with target populations of between
1000 and 2000, and a chief town at Emmeloord with a target population of 10,000.
Despite this careful planning, a number of problems have cropped up. The growth of
mechanized agriculture has resulted in fewer job opportunities than were originally
estimated. Paradoxically, Emmeloord has grown more rapidly than was expected,
whereas seven of the villages have populations of between 500 and 600, six of these
having recorded a decline in population since 1970.

The decline in rural population has been attributed to socio-economic changes
following the Second World War. With increasing affluence and personal mobility,
higher demands have been made on the provision of services: people in the rural areas
demand more from life and want not only more but also better facilities, and unless
villages are sufficiently large, they cannot cater for the demands made upon them.

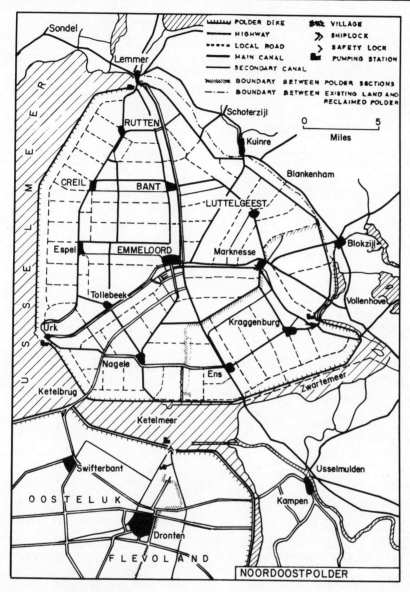

Figure 1.30 North East Polder
Reprinted with permission from *The Zuider Zee Guide* (1975) published by the Information and
Documentation Centre for the Geography of the Netherlands, figure 24.

2 What K system does the original plan (Figure 1.29) closely resemble?

3 Suggest ways in which the Dutch authorities could reverse the population decline in
the villages and increase the population of the area as a whole.

By 1957, when the Eastern Flevoland had been drained, the planners had gained
from earlier experience in the Wieringermeer and North East polders. Initially, the

settlement plan was similar to that of the North East Polder: there were to be 10 so called "A" centres having a local service function surrounding a single "B" centre (Dronten) which was to have a district service junction. A "C" centre (Lelystad) with a regional function was planned at the function of the four polders (Figure 1.26), but in the western corner of Eastern Flevoland.

4 Examine Figure 1.31 and, bearing in mind the problems in the other polders and the information shown in Figure 1.32, suggest why the planners changed their minds.

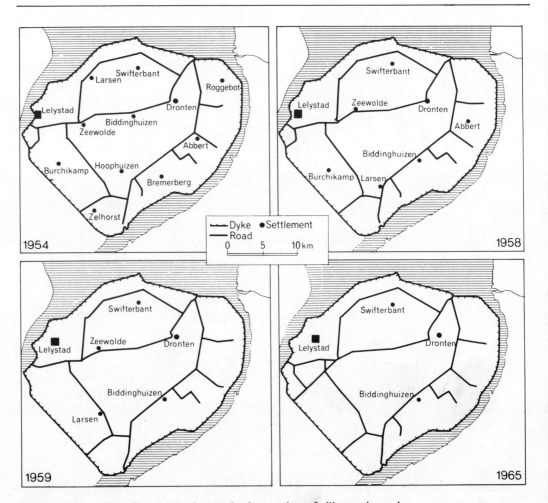

Figure 1.31 East Flevoland: changes in the number of villages planned
Reprinted with permission from *Village, Town, and City* by A. Adams and S. Dunlop (1976) published by Heinemann Educational Books, figure 7.9.

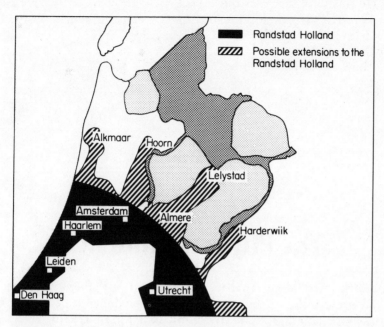

Figure 1.32   Outwards directed growth of the northern wing of the Randstad Holland
Reprinted with permission from *The Zuider Zee Guide* (1975) published by the Information and
Documentation Centre for the Geography of the Netherlands, figure 27.

The Physical Influence

It was stated in the introduction that the basic unit of settlement which recurs throughout the world is the village, the siting of which depends partly on the behavioural patterns of the community, partly on the location of existing villages and partly on the physical constraints of the environment. Prior to their specialization, many villages were sited where all essential resources — water, fuel, etc. — were accessible and, in societies with primitive technology, physical obstacles could not have been readily surmounted. But this is not to suggest that the location of settlements was initially dictated solely by physical constraints. A whole host of factors, including the need for defence, may have played a part. Furthermore, all constraints may change with time: the availability of arable and grazing land may alter as land becomes less productive; fuel resources and building materials may become exhausted or other sources found. Nonetheless, sites providing basic requirements — springs for example in areas devoid of surface drainage and raised terraces in areas liable to flooding would have been particularly attractive to pre-industrial settlers. In cases such as these, physical factors seem to have had an influence in settlement location.

PHYSICAL INFLUENCES ON SETTLEMENT SITING: A SIMULATION

This exercise illustrates the apparent influence of physical factors on the distribution of population and hence settlement location. Examine Figure 1.33, a hexagonally gridded map of an island, and Table 1.14 which provides information about physical factors in each of the cells. The island is to be settled by man in two phases — an agricultural one and an industrial one.

Phase One — Agricultural

The purpose of this first part of the exercise is to work out where 50,000 people are most likely to settle on the island in a pre-industrial period.

a Working in small groups, discuss the factors that are likely to influence the pre-industrial settlement patterns and allocate weights (points) to each factor listed in columns 2, 4, 6, and 8 of Table 1.14. For example, you may consider proximity to fresh water an important factor and allocate to this a positive weight, whereas you may consider marshland a repellent factor and award this a negative weight.

b Enter your weights in columns 3, 5, 7, and 9.

c Sum your weights for each cell and enter the totals in column 10.

In the absence of physical constraints on the island, all sites would be equally likely to be settled — which sites were actually settled would be a question of chance. This same chance element can be used in conjunction with the weightings for physical influences to find out where people are likely to settle on the island. The procedure is to give each cell a sequence of random numbers. The length of the random number sequence is determined by the total weight score of the cell. For example, if the sum of the total scores does not exceed a thousand, the first five cells might appear as in Table 1.12 where in cell 1, with a total weight score of +5, the random number sequence runs from 0 to 4; in cell 2, with a total weight score of +3, from 5 to 7; and so on. Notice that negative total weight scores are not given random number sequences — the cells they correspond to cannot be settled. To settle the island, a

Table 1.12   Example values

Col. 1	Col. 10	Col. 11	Col. 12
Cell	Total score	Random numbers	Number of times cell selected (max. 8)
1	+5	000–004	
2	+3	005–007	√√
3	−1	———	
4	+1	008–	√
5	+	009–011	√

table of random numbers (for example, Table 1.13) is then used. In Table 1.12, the random numbers are bracketed into groups of three to give numbers in the range

0 to 999. The first bracketed number, 35, lies outside the range of random number sequences given in column 11 of Table 1.12, though it would not if data for all the cells were included. The same is the case with the second bracketed number, 642. The third bracketed number, 7, corresponds to cell 2 in Table 1.12 and so a tick is entered in column 12 to indicate an allocation of 1000 people to that cell.

Table 1.13   Example random values

03	56	42	00	76	11	71	60	29	95
66	23	00	86	47	32	46	26	05	01
15	00	63	17	34	52	66	18	94	73

A set of random numbers is provided for you (Table 1.15).

2   The chance factor in settling the island is introduced by selecting random numbers from the random number table and hence the cells that correspond to those numbers. You are required to make 50 allocations in order to settle 50,000 people (50 x 1000). Record the number of allocations in column 12, with a maximum density allowed per cell of 8000 people, that is, eight ticks.

Phase Two — Industrial

Railways have been built; light industry has developed in areas already settled; the area has become a popular tourist resort, with passenger services to the mainland.

3   Another 50,000 people are to be settled in this phase, but since economic factors are now important in settlement siting you must reconsider your weightings. For example, proximity to existing settlements may be very attractive, especially where railways have been constructed, and so may areas of high tourist potential.

   a Discuss these factors in your groups and decide upon appropriate weights which should be put in columns 3, 5, 7, and 9 of Table 1.16. You will also need to reconsider the relevance of the altitude and adjust weights accordingly. (See notes at the foot of Table 1.16.)

   b Complete column 10 of Table 1.16 and use the table of random numbers (Table 1.15) to allocate another 50,000 people. Remember that each cell can only be selected eight times, including Phase One selections.

4   With the aid of the blank hexagonal outline (Figure 1.34) construct a dot map to show the spatial variation of population density on this island. Allow one dot to represent 250 people. Spread your dots evenly within each cell.

5   Compare your completed map with Figure 1.35 and suggest reasons for similarities and differences. You may need to refer to the relevant Ordnance Survey sheet.

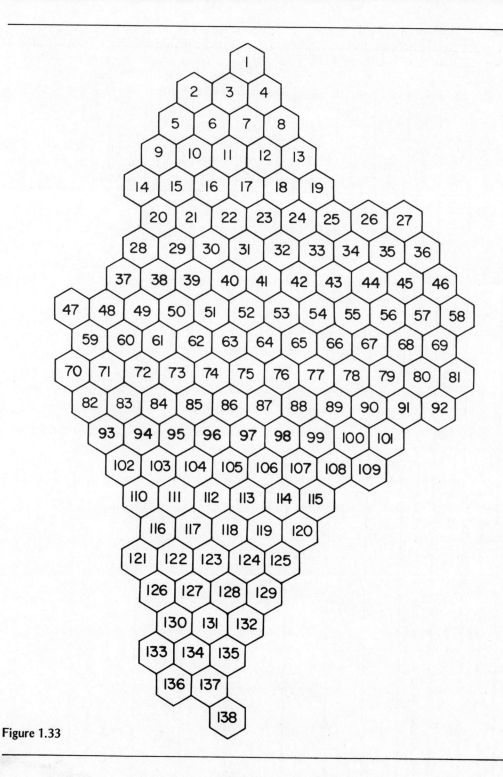

Figure 1.33

Table 1.14

											Column number
1.	2.	3.	4.	5.	6.	7.	8.	9.	10.	11.	12.
Cell No.	Geology	Weight	Altitude	Weight	River	Weight	Marsh (M) Uncliffed Coast (UC)	Weight	Total score	Random numbers	Number of times cell selected (max 8)
1	3		1				UC			000—	
2	3		2				UC				
3	3		1		NR		M				
4	3		4								
5	3		2								
6	3		1								
7	3		1		NR		M				
8	1		1		R		UC				
9	3		1		R						
10	3		1		R						
11	3		2								
12	1		1		R						
13	3		1				UC				
14	3		2								
15	3		2								
16	3		2		R						
17	1		3								
18	2		1		R		M				
19	2		1		R		UC				
20	3		2				UC				
21	3		1								
22	3		3								
23	2		1								
24	2		2								
25	2		2		R		UC				
26	2		3				UC				
27	1		4								
28	3		1		E		UC				
29	3		1								
30	3		2								
31	2		2								
32	2		1								
33	2		2								
34	2		2								
35	1		4								
36	1		4								
37	3		1		R		M/UC				

Table 1.14   Continued

											Column number
1.	2.	3.	4.	5.	6.	7.	8.	9.	10.	11.	12.
Cell No.	Geology	Weight	Altitude	Weight	River	Weight	Marsh (M) Uncliffed Coast (UC)	Weight	Total score	Random numbers	Number of times cell selected (max 8)
38	3		2								
39	3		3								
40	3		3								
41	2		2								
42	2		1		R						
43	2		2								
44	2		4								
45	1		4								
46	1		4								
47	2		2				UC				
48	2		2				UC				
49	3		1		R						
50	3		3								
51	3		3								
52	2		3								
53	2		1								
54	3		1								
55	2		2								
56	2		3								
57	1		4								
58	3		4								
59	3		1		E						
60	3		1		E						
61	3		1		E						
62	3		1		LBP						
63	2		2		R						
64	2		2		R						
65	2		3								
66	2		2								
67	2		3								
68	2		3								
69	1		4								
70	2		2				UC				
71	2		2								
72	3		2								
73	3		1								
74	1		2								

Table 1.14   Continued

											Column number
1.	2.	3.	4.	5.	6.	7.	8.	9.	10.	11.	12.
Cell No.	Geology	Weight	Altitude	Weight	River	Weight	Marsh (M) Uncliffed Coast (UC)	Weight	Total score	Random numbers	Number of times cell selected (max 8)
75	2		2								
76	1		2		R						
77	2		2		R R						
78	2		2		R						
79	2		3								
80	1		4								
81	1		4								
82	3		2		R		UC				
83	3		2		R						
84	3		2								
85	3		3								
86	1		3		R						
87	1		4								
88	1		4								
89	2		3								
90	2		3								
91	2		3								
92	2		3								
93	3		1		R		UC				
94	3		1								
95	3		2								
96	3		4								
97	1		4								
98	1		4								
99	1		3								
100	2		3								
101	2		2								
102	3		1				UC				
103	3		1		R		M				
104	3		1		R						
105	1		4								
106	1		4								
107	1		4								
108	2		2		R						
109	3		2								
110	3		1		E R		M				
111	3		1		R						

Table 1.14 Continued

| | | | | | | | Column number | | | | |
1.	2.	3.	4.	5.	6.	7.	8.	9.	10.	11.	12.
Cell No.	Geology	Weight	Altitude	Weight	River	Weight	Marsh (M) Uncliffed Coast (UC)	Weight	Total score	Random numbers	Number of times cell selected (max 8)
112	3		2								
113	1		4								
114	2		4								
115	3		2								
116	3		1		R		M				
117	3		1		R						
118	3		2								
119	1		4								
120	3		2								
121	3		2		R		UC				
122	3		1								
123	3		2								
124	1		4								
125	2		3								
126	3		1				UC				
127	3		2		R						
128	3		4								
129	3		2								
130	3		1		R		UC				
131	3		2								
132	1		2								
133	3		1				UC				
134	3		1		E						
135	3		1								
136	3		1				UC				
137	3		2								
138	3		2								

Key to columns:

2 Geology: 1 = chalk, 2 = sandstone and gravels, 3 = clay and alluvium

4 Altitude: 1 = 0 to 30 m, 2 = 31 to 60 m, 3 = 61 to 90 m, 4 = above 90 metres

6 Rivers: R = river, NR = navigable river, E = estuary, LBP = lowest bridging point

Table 1.15   Random numbers

60 93 52 03 44	87 19 54 60 92	73 79 64 57 53	12 30 28 07 83
60 97 09 34 33	30 34 24 02 77	98 52 01 77 67	76 37 84 16 05
71 43 00 49 09	47 32 46 26 05	98 88 46 62 09	76 35 59 37 79
18 74 39 24 23	75 23 76 20 47	69 23 46 14 06	71 58 66 34 17
29 40 52 42 01	49 13 90 64 41	11 80 50 54 31	85 49 65 75 60
60 79 01 81 57	57 17 86 57 62	83 45 29 96 34	15 29 27 61 39
78 21 21 69 93	35 90 29 13 86	19 56 54 14 30	52 30 87 77 62
03 99 11 04 61	93 71 61 68 94	38 55 59 55 54	21 70 67 00 01
03 55 19 00 70	09 48 39 40 50	88 68 54 02 00	94 53 89 11 43
63 57 33 21 35	05 32 54 70 48	99 59 46 73 48	95 29 40 05 56
94 55 93 75 59	49 67 85 31 19	41 84 98 45 47	66 98 63 40 99
41 61 57 03 60	64 11 45 86 60	46 35 23 30 49	80 62 05 17 90
69 91 62 68 03	66 25 22 91 48	11 08 79 62 94	76 62 11 39 90
09 89 32 05 05	14 22 56 85 14	45 15 51 49 38	96 29 77 88 22
19 52 35 95 15	65 12 25 96 59	52 70 10 83 37	69 57 21 37 98
67 24 55 26 70	35 58 31 65 63	57 27 53 68 98	76 46 33 42 22
05 24 92 93 29	19 71 59 40 82	65 48 11 76 74	43 70 86 63 54
60 58 44 73 77	07 50 03 79 92	17 54 67 37 04	29 76 08 36 37
19 50 23 71 14	69 97 92 02 88	98 38 03 62 69	74 28 77 52 51
91 49 91 45 23	68 47 92 76 86	94 86 43 19 94	94 75 08 99 23
80 33 69 45 98	26 94 03 68 58	52 56 76 43 50	53 14 03 33 40
21 81 85 93 13	93 27 88 17 57	78 49 89 08 30	07 08 28 50 46
44 10 48 19 49	85 15 74 79 54	90 76 70 42 35	57 60 04 08 81
50 27 39 31 13	41 79 48 68 61	40 18 82 81 93	55 41 18 56 67
41 39 68 05 04	90 67 00 82 89	80 12 43 56 35	95 08 30 67 83
88 49 29 93 82	14 45 40 45 04	34 41 48 21 57	74 84 39 34 13
51 47 46 64 99	68 10 72 36 21	63 43 97 53 63	42 83 60 91 91
53 85 34 13 77	36 06 69 48 50	32 64 25 28 61	49 36 47 33 31
56 46 39 93 80	38 79 38 57 74	50 55 61 76 95	46 06 22 76 47
96 29 63 31 21	54 19 63 41 08	55 73 25 62 34	71 17 11 51 02
24 63 73 87 36	74 38 48 93 42	65 13 85 68 06	12 35 91 86 01
30 93 44 77 44	07 48 18 38 29	38 00 10 21 76	25 89 12 04 05
08 42 26 89 53	19 64 50 93 03	69 57 26 87 77	15 95 33 47 64
99 01 90 25 29	09 37 67 07 15	24 12 26 65 91	88 67 67 43 97
12 80 79 99 70	80 15 73 61 47	67 04 90 90 70	98 95 11 68 77
66 06 57 47 17	34 07 27 68 50	79 49 50 41 46	63 81 33 98 85
31 06 01 08 05	45 57 18 24 06	91 70 43 05 52	56 29 90 74 39
22 88 84 88 93	27 49 99 87 48	20 05 77 31 56	74 62 28 46 70
83 08 01 24 51	38 99 22 28 15	15 53 33 49 24	97 37 62 75 85
19 61 27 84 30	11 66 19 47 70	98 08 62 48 26	86 86 81 26 65
39 14 17 74 00	28 00 06 43 38	92 69 44 82 97	31 34 02 62 56
85 26 97 76 02	02 05 16 56 92	33 18 51 62 32	33 09 38 52 47
64 75 68 04 57	08 74 71 28 36	27 36 98 66 02	03 27 44 34 23
16 44 42 43 34	36 15 19 90 73	59 06 67 59 74	85 13 03 25 52
92 90 15 18 78	56 44 12 29 98	49 71 29 73 80	06 45 32 53 11
12 88 39 73 43	65 02 76 11 84	58 27 56 17 64	45 81 95 29 79
72 46 13 32 30	21 52 95 34 24	89 51 41 17 88	55 74 11 40 14
79 78 22 39 24	49 44 03 04 32	37 40 29 63 97	96 60 95 82 32
35 33 77 45 38	44 55 36 46 72	97 12 54 03 48	36 90 84 33 21
48 90 81 58 77	54 74 52 45 91	21 82 64 11 99	28 46 82 81 09

Table 1.15 Continued

04 28 50 13 92	03 52 96 47 78	35 80 83 42 82	21 03 29 10 50
92 58 10 22 62	14 90 56 86 07	22 10 94 05 58	71 47 94 50 27
81 07 73 15 43	06 83 05 36 56	14 66 35 63 46	83 21 05 14 66
90 96 04 18 49	20 11 74 52 04	15 95 66 00 00	68 74 99 51 48
35 70 00 47 54	39 80 82 77 32	50 72 56 82 48	41 46 88 51 49
12 47 05 65 00	06 28 89 80 83	13 74 67 00 78	18 47 54 06 10
13 78 01 36 32	01 75 87 53 79	40 41 92 15 85	66 67 43 68 06
26 74 30 53 06	32 88 65 97 80	08 35 56 08 60	29 73 54 77 62
96 75 00 90 24	86 50 75 84 01	36 76 66 79 51	90 36 47 64 93
14 92 43 96 50	87 51 76 49 69	91 82 60 89 78	93 78 56 13 68
70 31 20 56 82	46 85 05 23 26	34 67 75 83 00	74 91 06 43 45
90 85 06 46 18	69 24 89 34 60	45 30 50 75 21	61 31 83 18 55
96 93 68 72 03	14 01 33 17 92	59 74 76 72 77	76 50 33 45 13
46 42 75 67 88	19 47 60 72 46	43 66 79 45 43	59 04 79 00 33
86 28 36 82 58	56 30 38 73 15	16 52 06 96 76	11 65 49 98 93
79 24 68 66 86	81 30 44 85 85	68 65 22 73 76	92 85 25 58 66
14 73 88 66 67	17 46 85 09 50	58 04 77 69 74	73 03 95 71 86
45 13 42 65 29	92 05 24 62 15	55 12 12 92 81	59 07 60 79 36
55 21 02 97 73	60 01 40 72 01	62 44 84 63 85	42 17 58 83 50
46 16 28 35 54	36 16 81 08 51	34 88 88 15 53	01 54 03 54 56
70 29 73 41 35	16 31 55 39 69	80 39 58 11 14	54 35 86 45 78
05 68 67 31 56	25 95 59 92 36	43 28 69 10 64	99 96 99 51 44
32 97 92 65 75	13 57 44 72 00	69 90 26 37 43	78 46 42 25 01
24 78 18 96 83	29 59 38 86 27	94 97 21 15 98	62 09 53 67 97
40 90 20 50 69	17 72 70 80 15	45 31 82 23 74	21 11 57 82 53
20 09 49 89 77	86 88 75 50 87	19 15 20 00 23	17 97 41 50 77
94 04 99 13 45	44 98 91 68 22	36 02 40 08 67	78 43 86 62 76
58 83 87 38 59	95 81 90 68 31	00 91 19 89 36	95 21 66 48 65
19 05 61 39 39	58 07 26 89 90	60 32 99 59 55	93 86 54 46 08
75 81 48 59 86	17 19 59 61 31	10 12 39 16 22	83 82 45 26 92
52 62 30 79 92	87 88 52 61 34	31 36 58 61 45	87 52 10 69 73
78 78 80 65 33	81 71 91 17 11	71 60 29 29 37	74 21 96 40 49
23 20 90 25 60	39 51 03 59 05	14 06 04 06 19	29 54 96 96 16
28 31 13 11 65	27 69 90 64 94	14 84 54 66 72	61 95 87 71 00
64 03 23 66 53	93 39 94 55 47	94 45 87 42 84	05 04 14 98 07
36 69 73 61 70	52 16 29 02 86	54 15 83 42 43	46 97 83 54 82
35 30 34 26 14	04 73 72 10 31	75 05 19 30 29	47 66 56 43 82
60 53 04 51 28	70 25 42 43 26	79 37 59 52 20	01 15 96 22 67
07 75 95 17 77	90 41 59 36 14	33 52 12 66 65	55 82 34 76 41
77 60 36 56 69	45 24 02 84 04	44 99 90 88 96	39 09 47 34 07
73 25 87 17 94	39 00 40 21 15	59 58 94 90 67	66 82 14 15 75
68 66 57 48 18	41 94 15 09 49	89 43 54 85 81	88 69 54 19 94
03 46 95 06 78	53 47 30 75 41	53 63 37 08 63	03 74 81 28 22
27 49 37 09 39	63 33 52 04 83	43 51 43 74 81	58 27 82 69 67
29 71 83 84 47	10 40 45 54 52	34 03 06 07 26	75 21 11 02 71
11 16 17 85 76	97 58 65 47 16	50 25 94 63 45	13 05 81 62 18
44 37 21 54 86	68 22 42 34 17	73 95 97 61 45	76 16 05 74 11
66 08 32 46 63	01 30 47 75 86	56 27 11 00 86	09 08 85 03 95
45 83 81 81 35	87 08 33 14 17	21 81 53 92 50	94 89 77 86 36
90 55 35 75 48	47 14 33 40 72	64 63 88 59 02	49 55 41 79 94

62

Table 1.16

						Column number					
1	2	3	4	5	6	7	8	9	10	11	12
Cell number	Altitude	Weight	Railway	Weight	High Resort potential	Weight	Existing settlement	Weight	Total score	Random numbers	Number of times cell selected
1	1										
2	2										
3	1										
4	4										
5	2										
6	1										
7	1		✓								
8	1										
9	1		✓		✓						
10	1		✓								
11	2		✓								
12	1		✓								
13	1		✓								
14	2		✓		✓						
15	2										
16	2		✓								
17	3										
18	1		✓								
19	1		✓		✓						
20	2										
21	1		✓								
22	3										
23	1		✓								
24	2										
25	2		✓		✓						
26	3				✓						
27	4										
28	1										
29	1		✓								
30	2										
31	3										
32	3		✓								
33	2										
34	2		✓								
35	1										
36	1		✓		✓						
37	3										

Table 1.16   Continued

Column number											
1	2	3	4	5	6	7	8	9	10	11	12
Cell number	Altitude	Weight	Railway	Weight	High Resort potential	Weight	Existing settlement	Weight	Total score	Random numbers	Number of times cell selected
38	2		√								
39	2										
40	3										
41	2										
42	3		√								
43	2										
44	2		√								
45	1		√								
46	1										
47	3										
48	3										
49	2		√								
50	2										
51	2										
52	2										
53	2		√								
54	3		√								
55	3		√								
56	1										
57	1										
58	3		√								
59	2		√								
60	2		√								
61	2		√								
62	2		√								
63	3		√								
64	2		√								
65	2										
66	2										
67	2		√								
68	2		√								
69	1		√								
70					√						
71	2										
72	2										
73	2		√								
74	1		√								

Table 1.16   Continued

					Column number						
1	2	3	4	5	6	7	8	9	10	11	12
Cell number	Altitude	Weight	Railway	Weight	High Resort potential	Weight	Existing settlement	Weight	Total score	Random numbers	Number of times cell selected
75	2										
76	1										
77	2										
78	2										
79	2										
80	1										
81	1										
82	2										
83	2										
84	2										
85	2		√								
86	1										
87	1										
88	1										
89	2										
90	2										
91	2										
92	2										
93	2										
94	2										
95	2		√								
96	2										
97	1										
98	1										
99	1										
100	2										
101	2										
102	2										
103	2										
104	2		√								
105	1										
106	1										
107	1										
108	3										
109	3										
110	3										
111	2		√								

Table 1.16 Continued

					Column number						
1	2	3	4	5	6	7	8	9	10	11	12
Cell number	Altitude	Weight	Railway	Weight	High Resort potential	Weight	Existing settlement	Weight	Total score	Random numbers	Number of times cell selected
112	2										
113	1										
114	3										
115	3										
116	2										
117	2		√								
118	2										
119	1										
120	3										
121	2										
122	2		√								
123	2										
124	1										
125	2										
126	2		√								
127	2										
128	2										
129	3										
130	3										
131	2										
132	1										
133	2										
134	2										
135	2										
136	2										
137	2				√						
138	2										

Key to columns:

2 Altitude — same as before

4 Railway — tick denotes presence thereof

8 Existing settlement — transfer your results from column 12 of Table 1.12

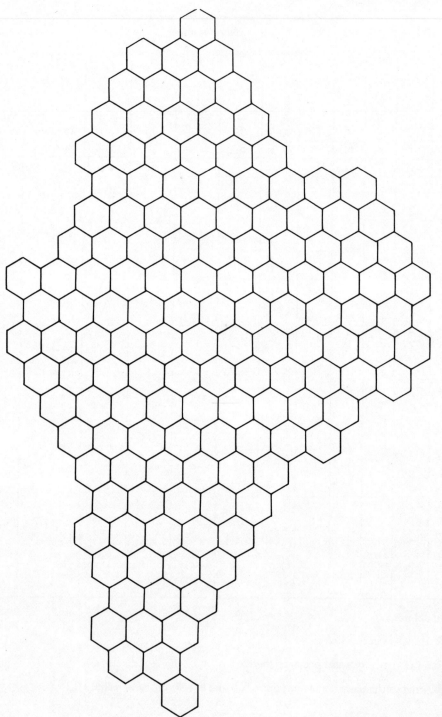

Figure 1.34

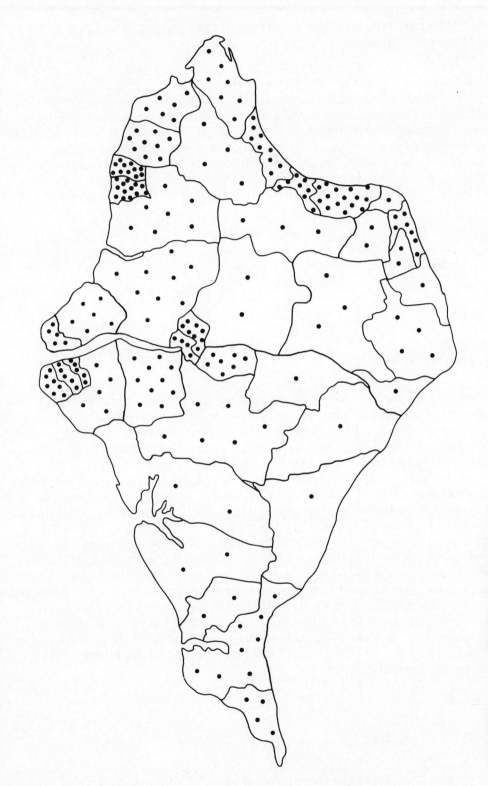

Figure 1.35   Isle of Wight: population density

SETTLEMENT IN THE ISLE OF WIGHT: THE INFLUENCE OF SLOPE, RELIEF AND GEOLOGY

The simulation exercise should have given you some insight into the effect of physical influences on settlement location and how economic factors can modify existing patterns. It has already been stated that most factors will change over time but, obviously, altitude and geology do not. What does change is man's response to altitude and geology. Nevertheless, the relative influence of altitude and geology on settlement siting can be evaluated. To show how physical factors can influence settlement, we shall examine the *possible* influence of land slope on settlement distribution. Average gradients were calculated for 138 equidistant points on the Isle of Wight, and the number of settlements associated with four slope classes noted (Table 1.17).

Table 1.17   Slope classes and settlement in Isle of Wight

Slope (in degrees)	Observed number of settlements (total 55)
Less than 3 (flat)	18
3 to 6 (gentle)	31
7 to 10 (moderate)	6
Greater than 10 (steep)	0

By examining the above data it could be inferred that gradient does influence the distribution of settlement. However, before we can come to a firm conclusion we ought to see if the apparent relationship between slope and settlements has arisen by chance. A suitable way of doing this is to apply a statistical test which will compare the observed frequencies of settlements in each slope category with the number of settlements which would be expected to be in each slope category if the distribution within each slope category were random. Firstly, we set up a null hypothesis (a negative proposition for the purpose of applying the statistical test) which states that the distribution of settlement is not related to gradient. If gradient does not influence the location of settlement, it would be reasonable to suppose that we should find an approximately even distribution of settlements irrespective of gradient. Therefore, in order to estimate the expected frequencies, we have to calculate the number of settlements we should expect to find in each slope category if they were distributed evenly. For example, if one-third of the study area were composed of moderate slopes, we should expect to find 18 settlements (one-third of 55) in the same area. Secondly, we test the probability that the observed frequencies are so different from the expected that they are unlikely to be due to chance variables by substituting the data in the chi-squared (χ^2) formula:

$$\chi^2 = \Sigma \frac{(O - E)^2}{E}$$

where O = observed frequencies
E = expected frequencies

The data is best tabulated and the value of chi-squared obtained as in Table 1.18.

Table 1.18   Chi-squared calculations

Slope (degrees)	Percentage of total area	O	E	$O-E$	$(O-E)^2$	$(O-E)^2/E$
Less than 3	45.8	18	25.2	−7.2	51.8	2.0
3 to 6	29.1	31	16.0	15.0	225.0	14.1
7 to 10	12.3	6	6.8	−0.8	0.6	0.1
More than 10	12.8	0	7.0	−7.0	49.0	7.0

Degrees of freedom = total number of slope categories less one   Total (Σ) = 23.2 (Answer)
          df. = 4 − 1 = 3

It is now possible to assess the extent to which the distribution we have been testing may be due to chance. The values 23.2 and 3 are the coordinates for use with the graph for the chi-squared test (Figure 1.36). By referring to the graph we see that the coordinate values in our example yield a probability value of less than 0.1 percent. In other words, we might expect the distribution to occur by chance less than 0.1 times

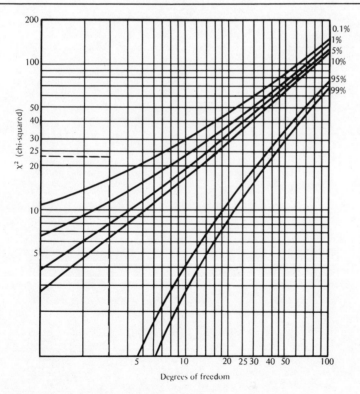

Figure 1.36   Graph for use in the chi-squared test

Reprinted with permission from *Statistical Methods and the Geographer* by S. Gregory (1978) published by Longman, London and New York

70

in 100; we can therefore reject the null hypothesis and accept that the distribution of settlement in the Isle of Wight is related to slope angle. What now needs to be considered is what such a relation means in geographical terms.

1 Make a tracing of Figure 1.37 which shows the distribution of settlements on the Isle of Wight and, with reference to Figures 1.38 and 1.39, test the influence of relief and geology on this distribution using data from Tables 1.19 and 1.20.

2 Formulate your null hypotheses.

3 Calculate the value of χ^2 and obtain the significance of your result from Figure 1.36.

4 Comment briefly on your results.

Table 1.19

Altitude	Percent total area
Less than 30 metres	36.3
31—60 metres	44.3
61—90 metres	7.0
More than 90 metres	12.0

Table 1.20

Geology	Percent total area
Chalk	22
Sandstones and gravels	47
Clay and alluvium	31

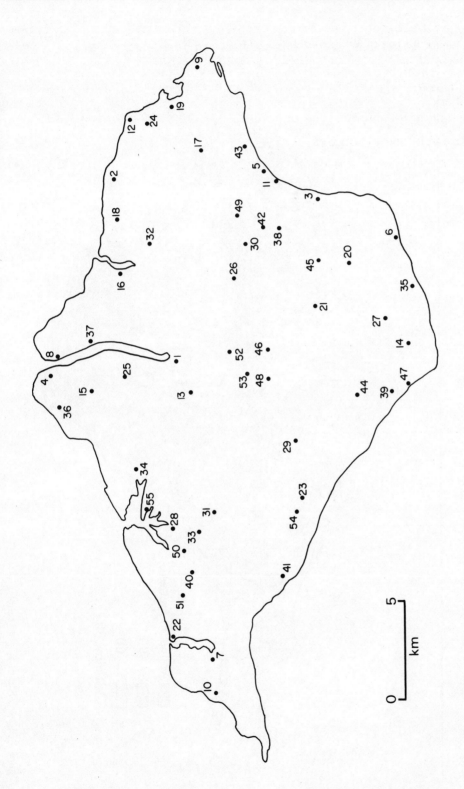

Figure 1.37 Isle of Wight: distribution of settlements

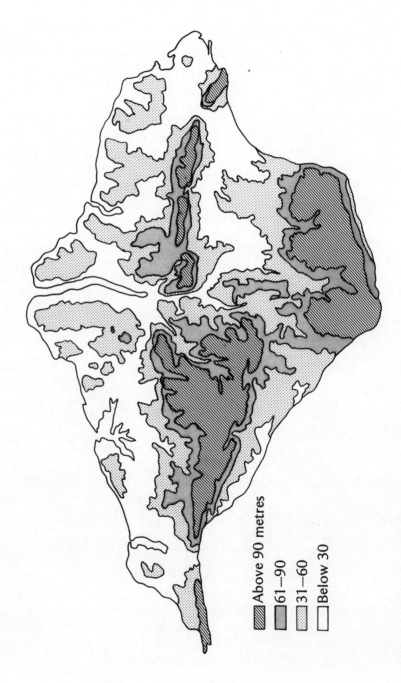

Above 90 metres

61–90

31–60

Below 30

Figure 1.38 Isle of Wight: relief

73

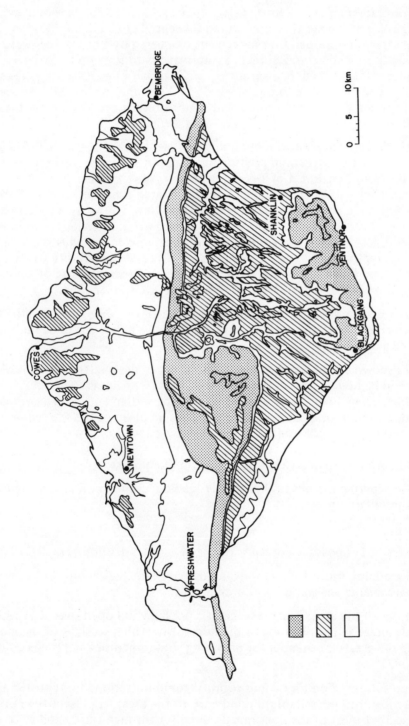

Figure 1.39   Isle of Wight: simplified geology

SITING A NEW TOWN

New towns are unique settlements since they are not natural developments of pre-existing villages; instead they are created by central government. Modern New Towns originate from the Garden City movement pioneered by Ebenezer Howard at the turn of the century, who envisaged them as alternatives to the unplanned, overcrowded and polluted cities which had sprung up in Victorian England during the industrial boom. Garden cities were to be spaciously laid out and surrounded by green belts which provided both farm produce and opportunity for recreation for the urban population.

In 1902, Howard succeeded in obtaining sufficient financial backing to start building his first city at Letchworth in Hertfordshire and in 1920 a second one at Welwyn. Under the New Towns Act of 1946, the postwar government, seeing how such towns could contribute to the rebuilding of the devastated urban environment and dispersal of industry, took control of their construction. New Towns were to have a number of specific objectives which included easing congestion in major metropolitan areas (overspill), providing improved living conditions in selected industrial areas and also providing employment in depressed areas by virtue of their ability to attract new industry. New Town legislation was later used to increase the size of a number of existing cities (expanded towns) in order to achieve similar aims.

Thus New Towns are quite different from other settlements since it is the Government which decides when they are to be built, where they are to be built and why they are to be built.

The area shown on the map (Figure 1.40) has been designated as one in which 75,000 people will be housed within the next few years. This aim is to be achieved by developing a new town *either* around an existing settlement *or* on a completely new site. Your initial task is to make a careful consideration of the site requirements before making a firm decision on the new town's location.

1   Overspill: Which urban area is this new town likely to serve?

2   Sites: Considering the physical character (geology, relief, drainage — Figure 1.41) assess the suitability of:
     i Misson
     ii Tickhill
     iii Bawtry . . . as possible cores for future new town development.

3   Land use competition: Discuss the land use conflicts that are likely to arise as a result of this area being designated for development.

4   Attractions of the area: The presence of woodland and open river sites would prove a big advantage in the planning of the new town. Which sections of the map extract indicate the greatest potential for providing these amenities and how could they be developed?

5   Communications: Good local and regional communications are regarded as essential in attracting both inhabitants and industry to the area. At present the railway stations are closed but most of the lines remain open for through traffic and to transport coal mined in the area.

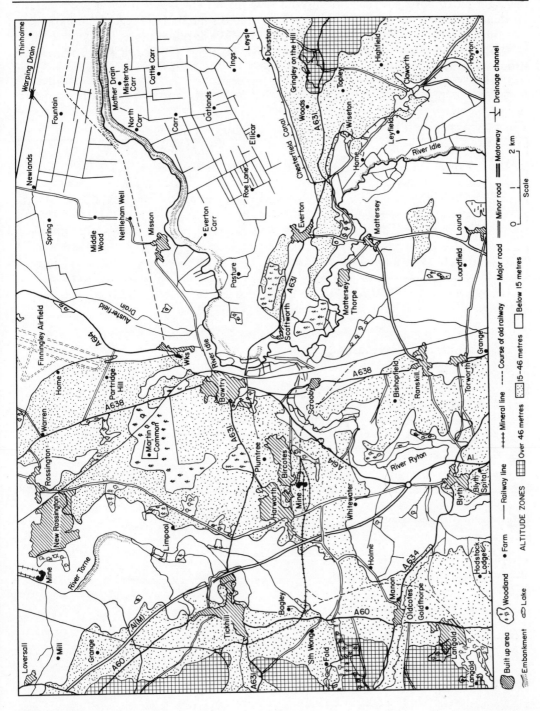

Figure 1.40

Having decided upon the location of your new town (which will cover approximately 16 square kilometres), with the aid of an annotated sketch map suggest and explain a possible reorganization of the transport links. Make as much use as possible of the existing network, bearing in mind the high costs involved of building new and long stretches of road and rail.

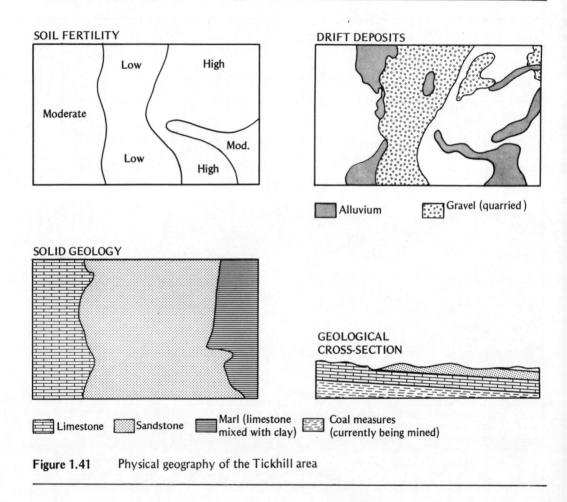

Figure 1.41   Physical geography of the Tickhill area

Conclusion

As J. H. Johnson (1972) pointed out, ". . . the distribution of towns and cities presents a confusing pattern which is difficult to sort out". However, by introducing a number of simplifying assumptions and by isolating important locational influences, Christaller and Lösch revealed a regular and logical pattern which lies behind the complex and seemingly chaotic arrangement of settlements found in all regions. Many properties of this regular pattern, such as average spacing between settlements, have been found to hold even in areas where population and resources are unevenly distributed. Moreover, the theories of Christaller and Lösch enable us not only to identify and describe regularities in the pattern of settlements, but also explain why the regularities exist and how they may have arisen. This is why theories of settlement distribution, and statistical techniques used to test them, have been invaluable to geographers.

Further Reading

Urban Geography: An Introductory Analysis, J. H. Johnson, Pergamon (1967), Chapter 5.

Human Geography: Theories and Their Applications, M. G. Bradford and W. A. Kent, Oxford University Press (1977), Chapters 1 and 4.

The Study of Urban Geography, H. Carter, Edward Arnold (1973), Chapter 5.

The North American City, M. H. Yeates and B. J. Garner, Harper & Row (1971), Chapter 6.

Introducing Towns and Cities, K. Briggs, University of London Press (1974).

Location in Space, P. Lloyd and P. Dicken, Harper & Row (1977), Chapter 3.

Pattern and Process in Human Geography, V. Tidswell, University Tutorial Press (1976), Chapters 10 and 11.

Geography: A Modern Synthesis, P. Haggett, Harper & Row (1975), Chapter 14.

Geography of Market Centres and Retail Distribution, B. J. L. Berry, Prentice-Hall (1967).

Settlement Patterns, J. Everson and B. FitzGerald, Longman (1971), Chapters 1 and 2.

Statistical Methods and the Geographer, S. Gregory, Longman (1971), Chapter 10.

Latin American Development, A. Gilbert, Penguin Books (1974).

CHAPTER TWO
INTERACTION BETWEEN SETTLEMENTS

Spheres of Influence

Man has always required land in order to satisfy basic needs such as food and water. Early settlements, and present-day ones of primitive tribes, required enough surrounding land to remain self-sufficient. The same relation between a settlement and its immediate surroundings, between a central place and its periphery, was characteristic of medieval and later towns. The people of the country exchanged food for the specialized services of the people of the town; the town was the centre of exchange. The catchment area of any particular town is known as the umland. In medieval England a farmer would be prepared to walk with his livestock about three miles to a market town to do his business; this means that markets were about six miles apart, a much closer spacing than present-day market towns whose umlands have expanded as a result of increased mobility.

An umland is also termed an urban field. As in a magnetic field, particles are attracted to a single point and, as in a gravitational field, planets are attracted to a star, so in an urban field people are "attracted" or look to a central town; in this sense the urban field or umland is a sphere of influence.

OBSERVED SPHERES OF INFLUENCE

1   The homes of consumers who travel to a particular retail centre can be plotted as a
dot map. This map will show the sphere of influence of a retail centre. Figure 2.1
shows where people live and where they shop in a central place hierarchy, lines joining
the two being known as desire or trip lines. From what we have discussed already,
suggest reasons for the differences between the two diagrams.

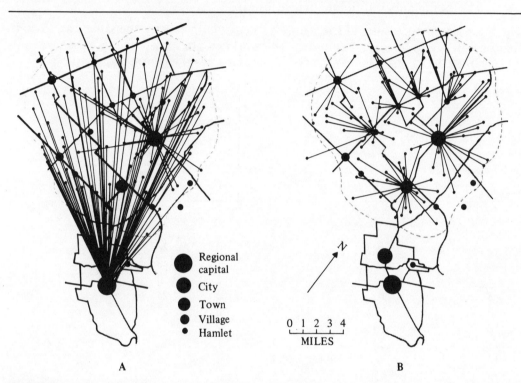

Figure 2.1   Desire lines
Reprinted with permission from "Cultural differences in consumer travel" by R. A. Murdie (1965),
Economic Geography, 41, figures 14 and 15.

2   A retail centre will not necessarily lie at the centre of gravity (mean centre) of its
catchment area. Figure 2.2 shows the distribution of a small sample of pupils' homes
that surround the school marked '■'. The centre of gravity of this distribution is
found by calculating the mean coordinate value of the x and y axes. In order to
calculate the mean coordinate of the x axis, the total number of points in each
column is multiplied by the mid-interval value. For example, two dots occur in the
column above the mid-interval value 2; hence the total 4. These totals are summed
and divided by the total number of points in the distribution (25). This may be
summarized in the formula:

$$\bar{X} = \frac{\Sigma xf}{n}$$

where $\bar{X}$ = mean coordinate value of the x axis
Σ = sum of
x = mid-interval value
f = number of points in each column
n = total number of points in distribution.

The mean coordinate value of the y axis is calculated in exactly the same way except that one uses rows instead of columns. The workings are shown in Table 2.1 and you will see that the intersection of the coordinates locates the centre of gravity.

Table 2.1

Coordinate x (columns)				Coordinate y (rows)		
mid-interval frequency				mid-interval frequency		
x	f	xf		y	f	yf
2	2	4		2	2	4
4	2	8		4	2	8
5	2	10		5	4	20
6	1	6		6	1	6
7	4	28	$n = 25$	7	2	14
8	2	16		8	3	24
9	1	9		9	3	21
10	3	30		11	2	22
11	2	22		12	1	12
14	3	42		14	4	56
15	3	45		15	1	15

$\Sigma xf = 220 \quad \bar{X} = \dfrac{220}{25} = 8.8$ $\qquad$ $\Sigma yf = 202 \quad \bar{Y} = \dfrac{202}{25} = 8.1$

In this example the mean centre is approximately 3.25 kilometres to the southeast. This eccentric location can be explained by the fact that the school as a whole draws upon a large number of boys from North London as well as more distant settlements such as St. Albans and Luton. In addition, some parts of the catchment area are considerably more accessible than others.

3 With reference to Figures 2.3 and 2.4, calculate the centres of gravity and compare the centrality of each centre with respect to the sample of consumers.

A measure of the dispersion of points around the mean centre is more accurately given by the Standard Distance which is calculated by using the formula

$$\text{Standard Distance} = \sqrt{\frac{\Sigma d^2}{n}}$$

where *d* = distance of any individual point from the mean centre.
  n = number points in the distribution.

If the points are distributed normally, approximately 65 percent of the points should lie within a radius of 1 standard distance from the mean centre.

4 For each distribution calculate the standard distance. Which centre appears to have the greater dispersion of consumers? What reasons can be put forward to explain these differences?

5 Comment on the size of sample.

From what we already understand about central places and the range of goods and services, it would seem reasonable to assume that with increasing distance from a central place, the fewer the consumers per unit area one is likely to attract. In Figure 2.5 a uniform distribution of consumers about a central place is pictured. To discover

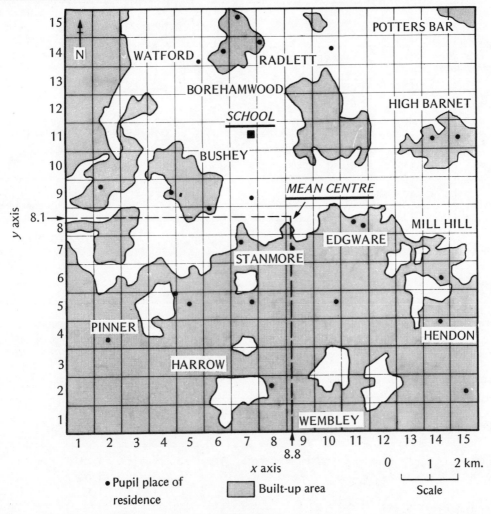

Figure 2.2   Distribution of sample of pupils' homes

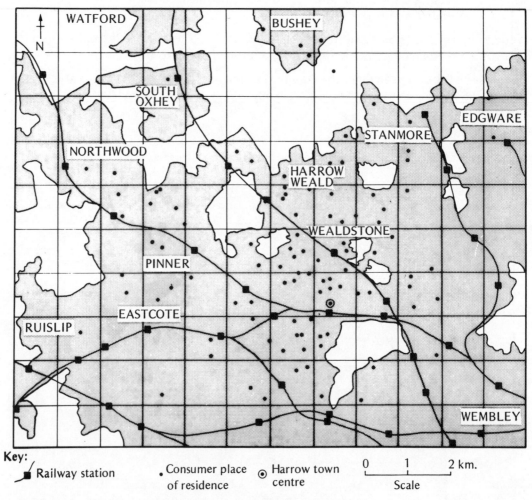

Key:

🔳 Railway station • Consumer place of residence ⊙ Harrow town centre

Scale 0 — 1 — 2 km.

Figure 2.3   Distribution of a sample of consumers' homes: Harrow

whether or not a density gradient exists, a series of concentric circles is constructed around the central place and the area of each one calculated. The workings are shown in Table 2.2 and the results plotted as a density gradient graph (Figure 2.6).

Table 2.2

Circle/ring	Area (km^2)		Number of consumers	Density (consumers per km^2)
1	$\pi1^2$	$= 3.14$	5	1.59
2	$\pi2^2 - 3.14 =$	9.4	16	1.70
3	$\pi3^2 - 9.4 =$	15.7	24	1.52
4	$\pi4^2 - 15.7 =$	22.0	24	1.09
5	$\pi5^2 - 22.0 =$	56.5	32	0.56

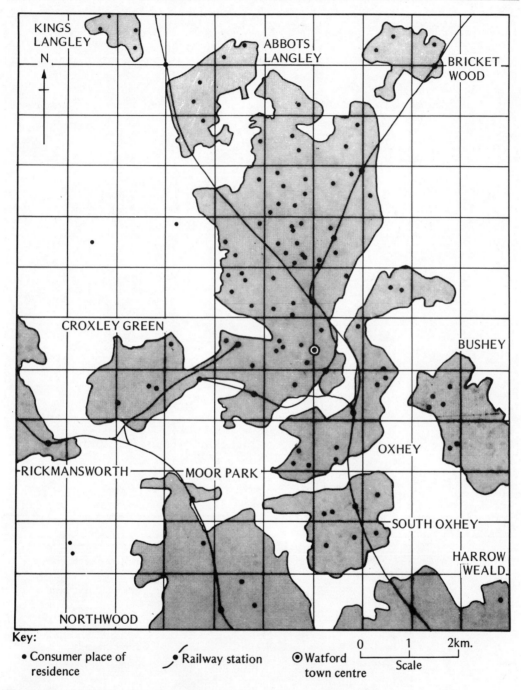

Key:

• Consumer place of residence

Railway station

⊙ Watford town centre

0   1   2km.
Scale

Figure 2.4   Distribution of a sample of consumers' homes: Watford

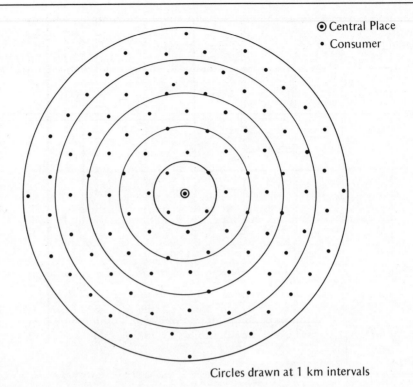

Circles drawn at 1 km intervals

Figure 2.5

6 With reference to Figures 2.3 and 2.4, construct graphs (Figure 2.7) to show the density gradients of consumers around Harrow and Watford. Again, can you suggest reasons for differences between the two areas? How do these results compare with those of the previous exercise?

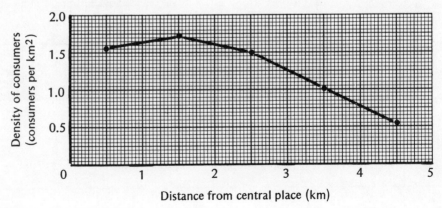

Figure 2.6

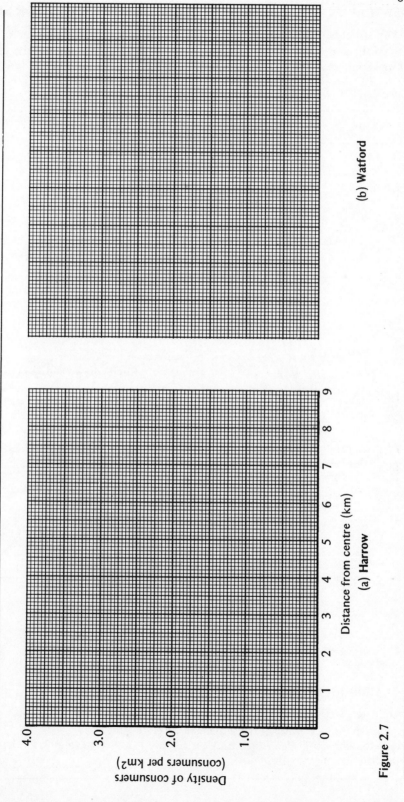

Density of consumers
(consumers per km2)

Distance from centre (km)

(a) Harrow

(b) Watford

Figure 2.7

NEWSPAPER CIRCULATION: AN ALTERNATIVE BASIS FOR SERVICE AREAS

Urban spheres of influence can be defined by many indicators, including shopping behaviour and commuting patterns. One important indicator is the circulation areas of newspapers and local newspaper reporting.

Table 2.3

Centre	Population
Norwich	121,800
Great Yarmouth	53,200
King's Lynn	32,100
Diss	5,400

1 With reference to Figure 2.8 and Table 2.3, can you suggest why there does not appear to be a perfect positive correlation between circulation area and population?

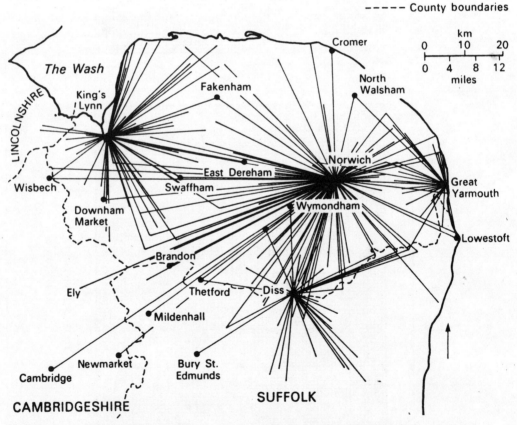

Figure 2.8   The distribution of local news reporting in Norfolk. The "trip" lines indicate a mention in the local paper of the settlement generating the line
Reprinted with permission from *Inside the City* by J. A. Everson and B. P. FitzGerald (1972) published by Longman, figure 104.

Ideally, a newspaper service area is based on calculations of the proportion a certain newspaper makes up of all newspapers received in each unit area. However, in this instance data are only available in the form of circulation/household ratios based on a sample survey.

2   Refer to Tables 2.4a and 2.4b, and on the outline map of the Harrow and Watford areas (Figure 2.9b) construct a choropleth map with at least four shades of intensity. Use a different colour for each newspaper.

3   To what extent are the service areas as depicted by newspaper circulation different given that the London boroughs are essentially suburban?

4   Are there any anomalies and if so what reasons can be put forward to explain them?

5   Does there appear to be a close correlation with degree of affluence? Using data from Tables 2.4a and 2.4b calculate the coefficient of correlation between newspaper circulation, one of the two variables, in each area and interpret your result. Why is it unwise, in this context, to consider all the wards?

Table 2.4a

Ward	Circulation/ household ratio (Harrow Observer)	*Percentage of households with >1 car	*Percentage of dwellings with all three amenities (hot water, bath, inside WC)
Pinner North	0.63	23.64	95.30
Harrow Weald	0.58	13.30	94.63
Stanmore North	0.51	23.37	93.45
Pinner South	0.59	13.03	97.57
Headstone	0.42	13.40	96.40
Wealdstone North	0.35	6.55	88.22
Wealdstone South	0.37	8.21	92.41
Kenton	0.33	13.06	97.35
Belmont	0.56	14.86	97.73
Stanmore South	0.57	8.32	94.19
Roxborne	0.41	9.82	96.98
West Harrow	0.50	6.84	92.93
Harrow on the Hill	0.60	10.83	87.06
Roxeth	0.08	8.20	95.66
Queensbury	0.17	8.67	96.56
Harefield	0.05		
Northwood	0.08		
Ruislip	0.03		
Eastcote	0.07		
South Ruislip	0.03		
Sudbury	0.08		
Preston	0.03		
Kenton Brent	0.07		
Kingsbury	0.11		

*Source: 1971 Census.

Table 2.4b

Ward	Circulation/ household ratio	*Percentage of households with >1 car	*Percentage of dwellings with all three amenities (hot water, bath, inside WC)
Leavesden South	0.30	11.09	97.47
Garston	0.42	11.21	96.33
St Andrews	0.63	22.72	93.08
Harebreaks	0.56	7.01	87.88
Bradshaw	0.32	4.17	69.40
Knutsford	0.45	8.59	97.52
Cassiobury	0.61	11.05	94.00
Queens	0.54	6.11	68.53
Harewoods	0.45	6.16	76.85
Kings	0.50	4.45	75.22
Oxhey	0.59	11.97	84.83
Mill	0.44	15.62	92.45
St James	0.48	16.05	97.58
Heath	0.33	22.62	92.82
Bovingdon	0.03		
Chipperfield	0.34		
Kings Langley	0.32		
Nashmills	0.44		
Bedmond	0.40		
St Stephen	0.08		
Flaunden	0.51		
Sarratt	0.47		
Langleybury	0.54		
Abbots Langley	0.63		
Leavesden North	0.53		
Aldenham West	0.09		
Chenies	0.06		
Chorleywood	0.07		
Rickmansworth	0.61		
Croxley Green North	0.46		
Croxley Green South	0.35		
West Hyde	0.58		
Northwood	0.04		
Moor Park	0.54		
Mill End	0.45		
Oxhey Hall	0.60		
Carpenter's Park	0.57		
Eastbury	0.19		
Blackwell	0.33		

*Source: 1971 Census.

R. E. Park and C. Newcomb (1933) pioneered research work on newspaper distribution areas in the USA; they constructed circulation boundaries through all places where at least 50 percent of the newspapers sold were distributed from a given centre (Figure 2.10).

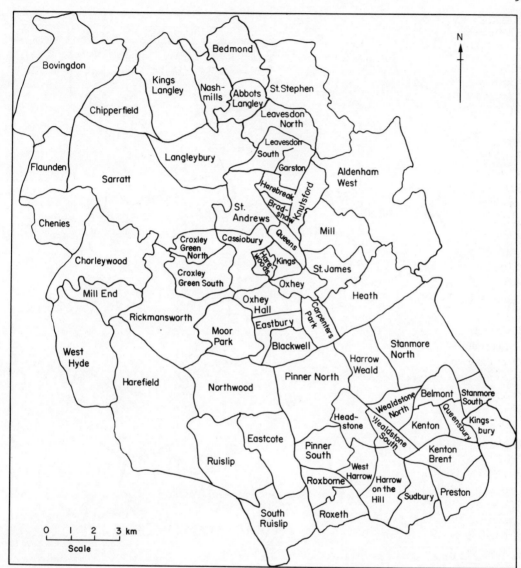

Figure 2.9a

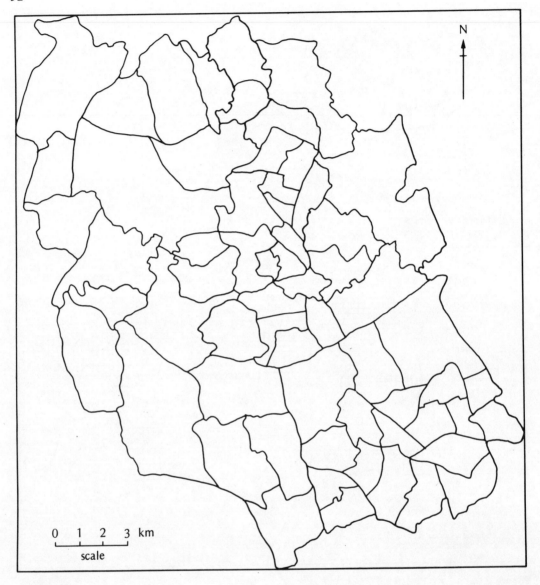

Figure 2.9b

6a Using data from Tables 2.4a and b, construct a transect graph (Figure 2.11) for both newspapers and label each curve.

  b Locate the circulation boundary between Watford and Harrow as depicted by newspaper circulation.

  c Attempt to explain why there does not appear to be an indifference zone as suggested by Park and Newcomb. A close examination of Figure 2.13 should be of some assistance.

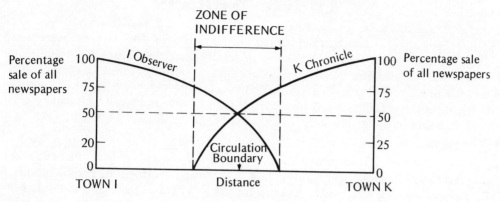

Figure 2.10

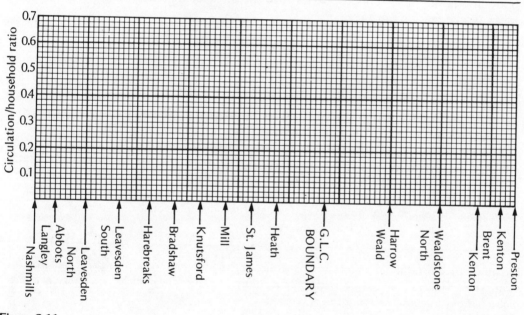

Figure 2.11

Gravity Models

The likeness between an urban field and a gravitational field has already been noted; it was explored by W. J. Reilly in his law of retail gravitation dating from 1929. Earlier work had also embodied gravity concepts. Stemming from the work of Reilly is a line of retail gravity models. Before studying these, we shall briefly consider another important and not unrelated line of gravity models which can be traced back to a paper by J. Q. Stewart, which came out in 1948, and a book by G. K. Zipf, published in 1949.

Quite independently, Stewart and Zipf arrived at the same idea. In a nutshell, and giving it a broad interpretation, this idea was that the movement of people, goods, or information between two towns depends on the size of the towns and the distance between them: the bigger the towns the greater the movement between them; the further the towns are apart, the less the movement between them. Specifically, the movement between two towns is directly proportional to the product of their populations, and inversely proportional to the distance between them. Writing these ideas as an equation we have

$$\text{Movement between town } i \text{ and town } j = K \frac{\left(\begin{array}{c}\text{Population} \\ \text{of town } i\end{array}\right) \times \left(\begin{array}{c}\text{Population} \\ \text{of town } j\end{array}\right)}{\left(\begin{array}{c}\text{Distance between} \\ \text{town } i \text{ and town } j\end{array}\right)^{\lambda}}$$

where K and λ are constants, that is values which remain fixed in a particular case. This expression is drawn directly from Isaac Newton's law of gravitation and it is therefore called a gravity model.* Measures of town size other than population could be used in the gravity model; any property of the town, such as employment or retail floorspace, which contributes to its attraction will do, and may be more suitable for a given case. Similarly, distance may be measured as the crow flies, or along transport routes, or may be more realistically measured as cost of travel or travel time.

---

*
Newton's law of gravitation states: every particle of matter attracts every other particle of matter with a force which varies directly as the product of the masses of the particles and inversely as the square of the distance between them; in symbols:

$$F = G\frac{M_i M_j}{d^2}$$

where F is the force exerted on two masses M_i and M_j at a distance d apart, and G is a universal constant known as the constant of gravitation. Thus two planets attract one another with a force determined by the gravitational law. The relationship, as we shall see, has been adapted to the attraction between towns and other social phenomena. It also has a direct significance in physical geography in producing tides.

The tides are caused by the attraction of the moon and of the sun on the earth. Figure 2.12 shows the influence of the moon on the oceans: it produces a high tide at the point A, nearest to the moon, and also at point B, farthest from it. The attraction by the moon on the earth is counterbalanced by an outwards or centrifugal force which tends to pull the two apart. At A the centri-

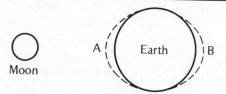

Figure 2.12

fugal force is weaker than the moon's attraction; at *B* the centrifugal force is stronger than the moon's attraction: in both cases the oceans bulge, in the former case at the point nearest to the moon and in the latter case at the point farthest from it. As the earth rotates, so the two high tides shift over the oceans. There are two high tides and two low tides every 24 hours and 50 minutes (the time taken for the moon to circle the meridian). The sun causes two-fifths of the tidal action and the moon three-fifths. At full and new moon, when the moon lies between the earth and the sun, the solar and lunar effects reinforce each other to produce the highest or spring tides; when the earth lies between the sun and the moon, the solar and lunar effects tend to cancel each other to produce the lowest or neap tides.

Gravity models of this kind have been fruitfully applied to a number of geographical problems. An interesting case is the Government's investigation into the Port of Bristol Authority's proposal for the building of nine new deep-water berths, two for bulk cargo and seven for general cargo, under Stage I of its Portbury project. The scheme was a large one costing £27 million at 1963 prices and, although in May 1965 the National Ports Council recommended to the then Minister of Transport that it should be approved, the Government decided to look further into the case. The Port of Bristol Authority's forecasts of overseas exports differed from the forecast made on its behalf by consultants; as a new set of data from a survey of the flow of goods to and from U.K. ports became available in 1964, an independent appraisal of the two forecasts by the government was possible.

The data used were for all goods handled by each of 25 ports, to and from 41 zones in the U.K. An initial analysis showed, somewhat surprisingly, that the amount of goods a port handles drops off rapidly with increasing distance from the port: the mean distance travelled for exports was 66 miles and for imports 36 miles; two-fifths of Britain's sea-going exports travel less than 25 miles, and two-thirds travel less than 75 miles. These results suggested that gravity model analysis might prove helpful in the investigation. An export and an import model were both built. The export model took the general form

$$\text{Exports moving from zone } i \text{ to port } j = K \frac{\left(\begin{array}{c}\text{Total exports} \\ \text{coming from zone } i\end{array}\right) \times \left(\begin{array}{c}\text{Total exports} \\ \text{handled by port } j\end{array}\right)}{(\text{Distance between zone } i \text{ and port } j)}$$

The model fitted the 1964 data reasonably well and some confidence could be attached to forecasts made by the model for 1980. The 1980 figure predicted for Bristol exports was 0.263 million tons; this compared with the consultants' figure of 0.465 million tons but was out of line with the Port of Bristol Authority's figure of 2.6 million tons which was the minimum figure needed to get a good rate of return on the investment from the Portbury development scheme. Even being as generous as possible with the tonnage of goods Bristol might hope to handle, the 1980 figure still fell short of the requisite 2.6 million by 42 percent. The model and related analysis thus made a strong case against the Port of Bristol Authority's figures and project which was in fact rejected.

PREDICTING SPHERES OF INFLUENCE: AN APPLICATION OF REILLY'S LAW

Reilly used gravity concepts in the delimitation of market areas. He assumed that the number of people a town attracts to its shops depends inversely on the distance between a person and the town and directly on the town's population. Given a choice of two towns at which to shop, a person will go to the town with the largest size/distance quotient since this town exerts the stronger "pull" on the person. Consider the following example. Town A is located between towns i and j as shown below:

```
          8 km                    10 km
 town  _____  town  _____  town
  i                        A                        j
(100,000)                               (75,000)
```

According to Reilly's model, the number of people living in town A and shopping in town i is

$$\frac{\text{Town } A\text{'s patronage}}{\text{to town } i} = \frac{\text{Population of town } i}{(\text{Distance from } A \text{ to } i)^2}$$

where we are using distance squared rather than straight distance. (The value of 2 is a constant but varies with type of shopping good. It will be discussed later.) The number of people living in town A and shopping in town j is

$$\frac{\text{Town } A\text{'s patronage}}{\text{to town } j} = \frac{\text{Population of town } j}{(\text{Distance from } A \text{ to } j)^2}$$

The relative proportion of people who live in town A and patronize the shops in each of towns i and j is given by the formula

$$\frac{\text{Town } A\text{'s patronage to town } i}{\text{Town } A\text{'s patronage to town } j} = \frac{(\text{Population of town } i/\text{distance from } A \text{ to } i)}{(\text{Population of town } j/\text{distance from } A \text{ to } j)}$$

which can be rearranged to produce

$$\frac{\text{Relative patronage from}}{\text{town } A \text{ to towns } i \text{ and } j} = \frac{\text{Population of town } i}{\text{Population of town } j} \times \frac{\text{Distance from } A \text{ to } j}{\text{Distance from } A \text{ to } i}$$

Using the data in the example we find

$$\frac{\text{Relative patronage from}}{\text{town } A \text{ to towns } i \text{ and } j} = \frac{100,000}{75,000} \times \frac{10^2}{8^2}$$

$$= 2.07$$

This means that twice as many shoppers from town A will visit town i than will visit town j.

The same kind of analysis could be applied to other forms of consumer behaviour. For instance, we could consider the competition for sales in town A of a local newspaper produced in town i and another local newspaper produced in town j. Given the same populations and distances as in the shopping example, the result would be the same but would mean that twice as many newspapers from town i are sold in town A than are sold there from town j.

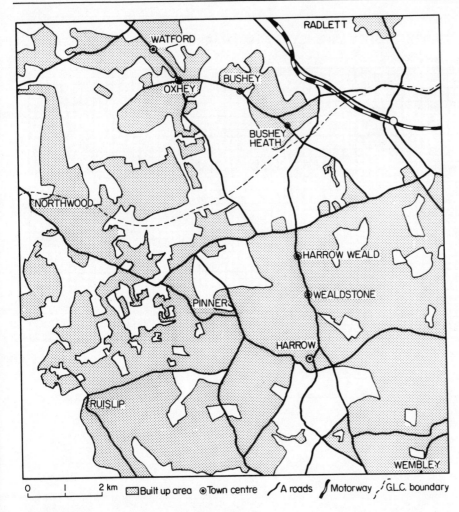

Figure 2.13

1 With reference to Figure 2.13 and Table 2.5, calculate patronage values for Harrow with respect to Oxhey, Bushey, Bushey Heath, Harrow Weald, and Wealdstone.

2 Plot your values on the section paper (Figure 2.14) and join the points.

3 How does this graph compare with the one you completed for the previous exercise? Suggest reasons for the differences.

Table 2.5

Centre	Population
Watford	77,800
Harrow	201,300

s—D

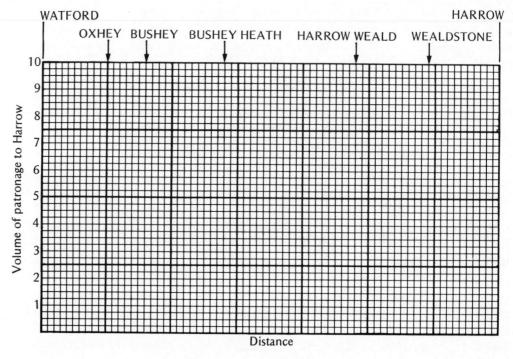

Figure 2.14

A MODIFICATION OF REILLY'S LAW

In 1949, P. Converse modified Reilly's law to predict the breaking point which demarcates a theoretical boundary between the two market areas. This may be calculated from the formula:

$$D_j = \frac{D_{ij}}{1 + \sqrt{\dfrac{P_i}{P_j}}}$$

where D_j = breaking point between town i in kilometres from j
D_{ij} = distance town i and j
P_i = population of town i (large)
P_j = population of town j (small)

1 With reference to Figure 2.13 and Table 2.5, calculate the breaking point between Watford and Harrow. Locate this boundary on Figure 2.13 and compare this result with your previous findings.

Converse's formula can also be used to demarcate the sphere of influence as illustrated in Figure 2.15. You will see that by connecting these break points you are in effect predicting that shoppers within this boundary would patronize centre A, whereas those outside would travel to the nearest alternative centre. This naturally assumes that the shoppers behave rationally, that is, patronizing the nearest shopping centre.

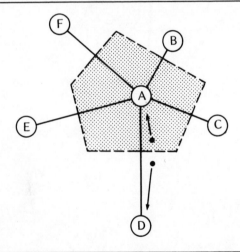

Figure 2.15

With reference to Figure 2.16:

Draw 4 straight lines radiating from Redbourn to Luton, Hemel Hempstead, St. Albans, and Harpenden and measure their lengths. There is no need to convert these lengths to kilometres.

b Using Converse's formula, calculate the break points between Redbourn and the four surrounding centres with the aid of population data from Table 2.6.

c Construct the market area of Redbourn as defined by Converse's formula.

Table 2.6

Centre	Population (000's)	Retail floorspace (thousands ft^2)	Road distance from Redbourn (km)	Average travel time (minutes)
Luton	166	884	10.2	11
Hemel Hempstead	69	361	9.5	10
St. Albans	52	402	6.7	7
Harpenden	24	108	5.3	6

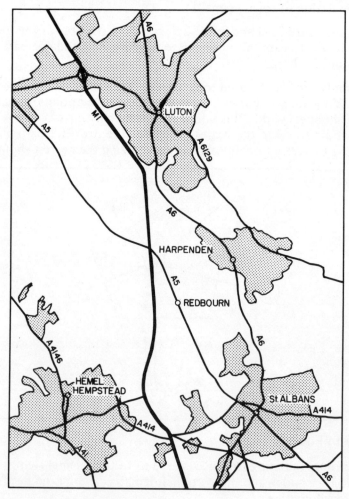

Figure 2.16   Study area: South Hertfordshire

REILLY'S MODEL: AN ALTERNATIVE APPROACH

Using Reilly's original formula, and retail floorspace data (Table 2.6), calculate patronage values for each of the alternative centres. Convert them to percentages and complete the percentage matrix (Table 2.7).

Worked Examples:

a *Redbourn patronage to Luton and St. Albans*

$$\frac{884}{402} \times \left(\frac{6.7}{10.2}\right)^2 = \frac{95}{100} \text{ or } 0.95$$

$$95 + 100 = 195$$

$$100 \div 195 = 0.51$$

Therefore 0.51 x 100 = 51 percent to St. Albans
0.51 x 95 = 49 percent to Luton.

b *Redbourn patronage to Hemel Hempstead and St. Albans*

$$\frac{402}{361} \times \left(\frac{9.5}{6.7}\right)^2 = \frac{223}{100} \text{ or } 2.23$$

$$223 + 100 = 323$$

$$100 \div 223 = 0.31$$

Therefore 0.31 x 100 = 31 percent to Hemel Hempstead
0.31 x 323 = 69 percent to St. Albans.

Table 2.7   Percentage matrix

	Luton	Hemel Hempstead	St. Albans	Harpenden	Total	Predicted percentage patronage
Luton			49			
Hemel Hempstead			31			
St. Albans	51	69				
Harpenden						
					Σ600	

Sum the rows of percentages and complete the total column. To adjust these percentages divide each total by 6.

These percentages are in effect the predicted values for Redbourn. A sample survey was conducted in Redbourn (1977) and the frequency of visits to each of the alternative centres are presented in Table 2.8 in the form of percentages.

Table 2.8   Sample survey results

Centre	Survey (percentages)
Hemel Hempstead	28.7
St. Albans	47.0
Luton	7.8
Harpenden	16.5

2   Compare the survey results with those you have predicted. How far do they diverge from the expected?

3   Can you suggest reasons for the differences in the results, bearing in mind that we used retail floorspace data for "mass" and the most direct physical route for "distance"? Are the shoppers in Redbourn likely to use the motorway to reach Luton? Luton is at a disadvantage in that there is no direct bus route and parking facilities are inadequate for the size of the town.

D. L. HUFF'S GRAVITY MODEL

No further progress was made with retail models until the work of D. L. Huff appeared in 1962. Huff's model, rather than merely fixing a breaking point between two towns, predicts the probability of consumer trips from several towns to a number of competing shopping centres. Expressed as an equation his model reads

$$
\begin{array}{l}
\text{Probability of a} \\
\text{consumer in town} \\
i \text{ shopping in} \\
\text{town } j
\end{array}
=
\dfrac{\dfrac{\text{Selling area for a class of goods in town } j}{\text{Travel time from town } i \text{ to town } j}}{\dfrac{\text{Total selling area for a class of goods in all towns being studied}}{\text{Total travel time from town } i \text{ to all other towns being studied}}}
$$

A symbolic version of this expression is given in the footnote.* The selling area (retail floorspace) of a class of goods is a measure of shopping centre size and may be used separately for different goods. Thus if we wish to calculate the probability of a consumer in town i shopping in town j for furniture, we would use retail floorspace devoted to furniture in the above formula.

Take the example of the four shopping centres in south Hertfordshire — Harpenden, Hemel Hempstead, Luton, and St. Albans. According to Huff's model, the probability of consumers from Redbourn (town i) travelling to Luton (town j) to purchase goods is

$$
p_{ij} = \dfrac{\dfrac{\text{Retail floorspace in Luton}}{\text{Travel time from Redbourn to Luton}}}{\dfrac{\text{Total retail floorspace in all four centres}}{\text{Total travel time to all four centres}}}
$$

---

*
In statistical notation Huff's formula reads

$$
P_{ij} = \dfrac{\dfrac{S_j}{T_{ij}^{\lambda}}}{\sum\limits_{j=1}^{n} \dfrac{S_j}{T_{ij}^{\lambda}}}
$$

where P_{ij} = probability of a consumer at point of origin i travelling to centre j
S_j = size of shopping centre
T_{ij} = journey time from i to j
λ = (lambda) is an exponent or constant. It is the power to which "distance" is raised and varies inversely with the order of a good. Consumers are prepared to travel long distances in order to purchase furniture; therefore lambda would be low.
$\sum\limits_{j=1}^{n}$ = a statistical shorthand instruction to add all values for the j centres, starting with $j = 1$ and ending with $j = n$, where n is the total number of j centres.

Using survey data we have

$$\frac{\dfrac{884}{11}}{\dfrac{884}{11} + \dfrac{361}{10} + \dfrac{402}{7} + \dfrac{108}{6}} = \frac{80.36}{191.89} = \underline{\underline{0.41}}$$

This is repeated for the remaining three centres.

Hemel Hempstead

$$\frac{\dfrac{361}{10}}{\dfrac{884}{11} + \dfrac{361}{10} + \dfrac{402}{7} + \dfrac{108}{6}} = \frac{36.1}{191.89} = \underline{\underline{0.19}}$$

St. Albans

$$\frac{\dfrac{402}{7}}{\dfrac{884}{11} + \dfrac{361}{10} + \dfrac{402}{7} + \dfrac{108}{6}} = \frac{57.43}{191.89} = \underline{\underline{0.30}}$$

Harpenden

$$\frac{\dfrac{108}{6}}{\dfrac{884}{11} + \dfrac{361}{10} + \dfrac{402}{7} + \dfrac{108}{6}} = \frac{18.0}{191.89} = \underline{\underline{0.10}}$$

The total probability comes to 1.0 as one would expect.

1   Converting these probabilities to percentages, for example, 0.41 becomes 41 percent, how do these predicted results compare with your previous findings about shopping behaviour in the area.

The Huff model can also be used to work out the actual number of consumers likely to shop at a given shopping centre. The formula is

Number of consumers   Total number
travelling from town   = of consumers $\times p_{ij}$
i to town j to shop   living in town i

The population of Redbourn is 3835 and the probability of travelling from Redbourn to St. Albans to shop is 0.3. Thus some 1151 people residing in Redbourn (3835 x 0.3) are likely to shop in St. Albans.

2   Calculate actual patronage figures from Redbourn to the other shopping centres. What factors may lead to these figures being too high?

DEFINING SPHERES OF INFLUENCE

Huff's model also shows the decline of attracting power of the various centres with distance, unlike Converse's formula which simply establishes the 0.5 probability line (breaking point) between two centres. The retail trading area is therefore depicted in the form of probability contours (Figure 2.17).

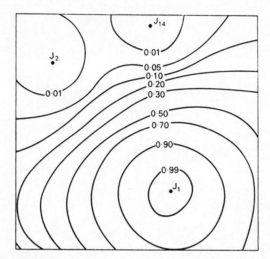

Figure 2.17   A retail trading area portrayed in terms of probability contours
Reprinted with permission from D. L. Huff, "A probabilistic analysis of shopping centre trade areas", *Land Economics*, Volume 39 (© 1963 by the Board of Regents of the University of Wisconsin System), figure 2.

Examine Figure 2.18 which shows a rural part of the Midlands. This area of approximately 500 square kilometres is bounded by the centres Nottingham in the north, Leicester in the south, Loughborough in the west and Melton Mowbray in the east. With the aid of Tables 2.9 and 2.10 calculate probabilities of consumer trips for footwear from each of the small settlements (i) to one of the main centres (j). These calculations as we have already seen are not complicated but lengthy. You are therefore advised to work in groups. For simplicity we shall assume that lambda is 1.0; therefore your calculations are similar to the ones completed for Redbourn in the previous exercise.

Example: Barton-in-Fabis (i centre) and Nottingham (j centre)

$$\frac{\dfrac{92000}{8}}{\dfrac{92000}{8}+\dfrac{91000}{30}+\dfrac{9000}{15}+\dfrac{6000}{28}}=\frac{11500}{15155}=\underline{\underline{0.76}}$$

Therefore the probability of a consumer at Barton-in-Fabis visiting Nottingham to buy shoes is 0.76. Next the probability value for Plumtree would be calculated. Note that these probabilities do not add up to 1.0 because we are only considering a single

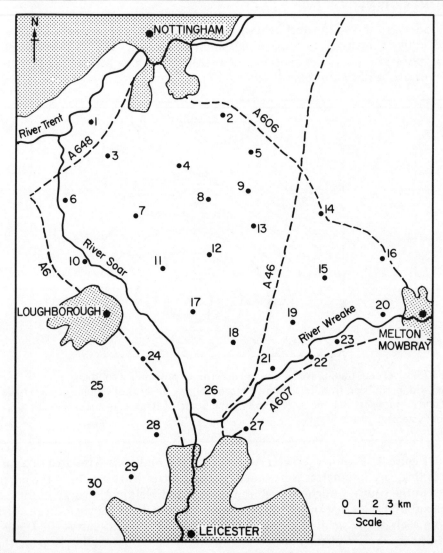

Figure 2.18   The distribution of settlements in part of the East Midlands

Table 2.9

j centres	Footwear retail floorspace (sq. ft.)
Nottingham	92,000
Leicester	91,000
Loughborough	9,000
Melton Mowbray	6,000

Source: Census of Distribution, 1971

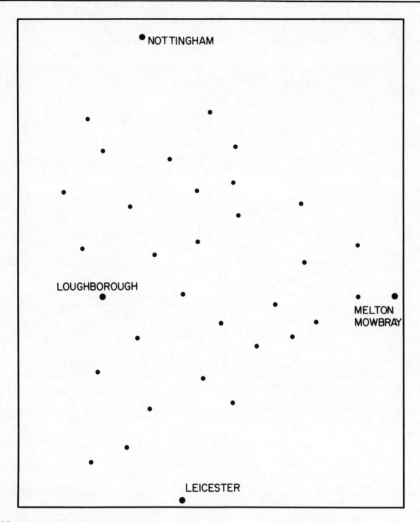

Figure 2.19

j centre. It is when we consider all the *j* centres that the total probability is 1.0. For example Barton-in-Fabis probability to Nottingham plus the probability to Leicester plus the probability to Loughborough plus the probability to Melton Mowbray equals 1.0.

2 Plot your probability values on Figure 2.19 and construct probability contours (isopleths) at intervals of 0.1.

3 Compare your predicted probability maps with the relevant Figure 2.20a, b, c, or d which is based on results obtained from a sample household survey conducted in 1976. Comment on your comparisons.

4 With reference to Figure 2.21 explain the relationship between (a) the size of (*j*) and (b) type of good and probability gradient.

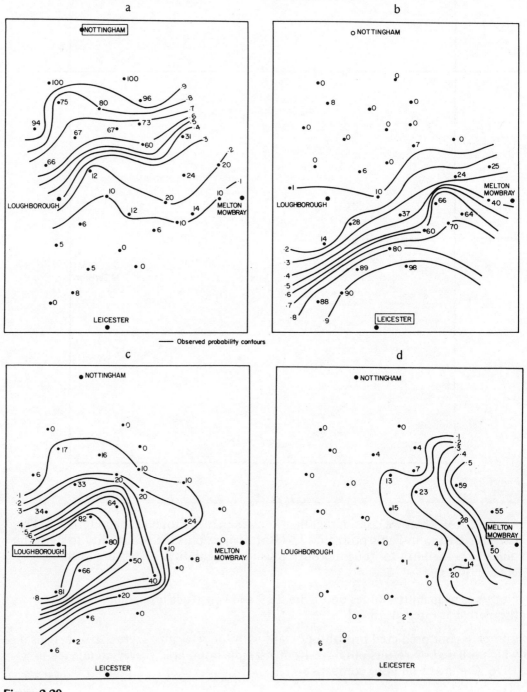

Figure 2.20

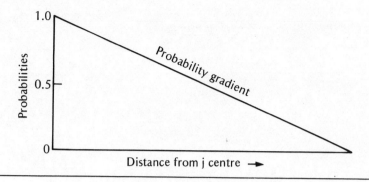

Figure 2.21

Table 2.10

Map Ref. No.	i centres	Estimated travel times to j centres (minutes)			
		Nottingham	Leicester	Loughborough	Melton Mowbray
1	Barton in Fabis	8	30	15	28
2	Plumtree	8	27	16	17
3	Gotham	11	28	13	24
4	Bunny	10	25	11	22
5	Stanton on the Wolds	12	25	16	17
6	Kingston on Soar	14	25	9	25
7	East Leake	16	23	8	19
8	Wysall	15	25	11	19
9	Widmerpool	17	24	14	17
10	Normanton on Soar	22	22	6	22
11	Hoton	16	19	4	17
12	Wymeswold	17	21	8	13
13	Willoughby on the Wolds	18	20	12	14
14	Upper Broughton	17	23	17	9
15	Grimston	23	21	16	8
16	Ab Nettleby	23	25	20	4
17	Walton on the Wolds	20	18	7	17
18	Seagrave	27	18	11	15
19	Ragdale	24	17	14	12
20	Asfordby Hill	27	18	23	2
21	Thrussington	28	12	14	10
22	Rotherby	30	12	21	7
23	Frisby on the Wreke	29	14	20	5
24	Quarndon	24	12	3	20
25	Woodhouse Eaves	28	13	7	24
26	Cossington	29	9	10	13
27	Syston	29	7	13	12
28	Cropston	31	8	12	18
29	Anstey	34	5	14	20
30	Groby	39	5	17	23

Territories and Boundaries

Allied to the idea of urban fields is the notion of territories and boundaries. The area under the control or influence of one settlement may end abruptly where it abuts the area belonging to an adjacent settlement. The border between the two may or may not be a closely guarded barrier. The parishes of England have not been the cause of major territorial disputes, though records exist of people from one parish "illicitly" acquiring some of the resources in a neighbouring parish. National boundaries, however, especially those in central Europe, have constantly migrated as a result of war.

Generally speaking, the area next to a boundary is less attractive to economic activity than the interior locations; this is known as the halo effect. In Russia, the growth of cities between 1913 and 1939 was about five times greater in cities over 150 miles from the western frontier than in those within 150 miles of the western frontier. This is the result of a deliberate Soviet government policy to develop new urban areas in regions such as Siberia and Central Asia which were far removed from the militarily vulnerable European border. Most state capitals, county seats, and the like are located in the centre of the territory they govern and administer, mainly so that they are accessible to all the people they serve. The relative remoteness of the Scots and Welsh from central British government in London is a contributory factor to the plea for devolution.

THE HALO EFFECT IN AFRICAN NATIONS

Figure 2.22 shows the location of national capitals in African countries.

a Using an atlas which has thematic maps (vegetation, population density etc) of Africa, consider how well some of the capitals are placed for administering the people they serve.

b Explain why some of the capitals are not centrally located.

A bounded region or territory has at least two interesting properties — size and shape. According to the theoretical schemes of Christaller and Lösch, a region should consist of a set of hexagons, every one the same size (see p. 17). This notion was challenged by W. Isard in 1956 who showed that in practice a regular pattern of hexagons is unlikely to occur. Population density is far higher in and near central places than in

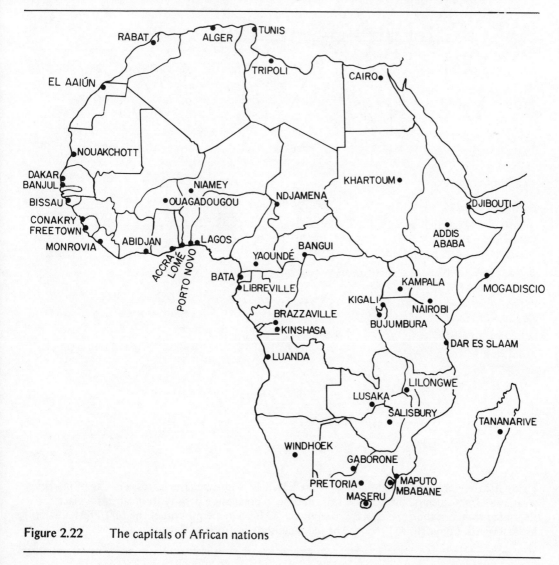

Figure 2.22   The capitals of African nations

112

rural areas. Assuming that market areas serve roughly the same number of people, the ones near and in higher-order central places will be smaller than those in peripheral areas. This effect arises because people tend to cluster or agglomerate and is therefore called the effect of agglomeration; it leads to a significant distortion of the Löschian landscape in which polygon size varies with population density (Figure 2.23).

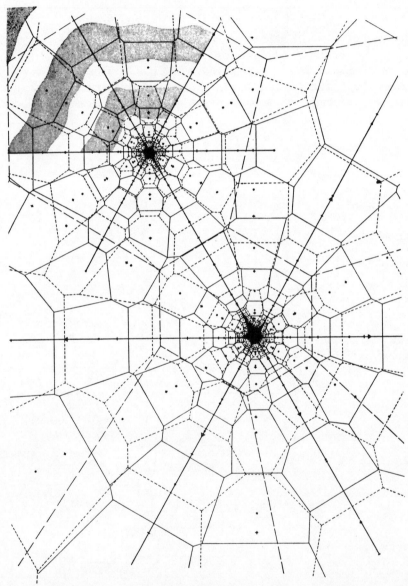

Figure 2.23   W. Isard's modification of the Löschian landscape. The increase in size of market areas away from a central place follows the decline in population density from a central place Reprinted from *Location and Space Economy* by W. Isard by permission of the MIT Press, Cambridge, Massachusetts. Copyright © 1956 by The Massachusetts Institute of Technology.

THE SIZE OF TERRITORIES

Figure 2.24 is a scattergraph of population density versus area for the states and territories of the United States in 1890. Rank correlating the two variables yields a correlation coefficient of −0.88, which is significant at the 99.9 percent level.

1 To what extent do the scattergraph and correlation analyses validate the hypothesis that size of territory and population density are related?

2 Consider how historical factors (such as when the states were first established), and physical factors (study an atlas) may influence the pattern shown in Figure 2.24.

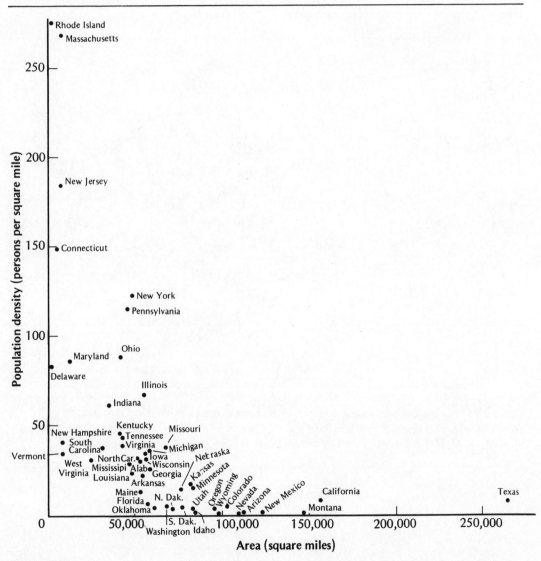

Figure 2.24 The relationship between population density and size of territory for the United States in 1890

114

The size of urban fields also varies with population density.

3   Figure 2.25 shows the circulation areas of metropolitan newspapers, and Figure 2.26 population density variations, for the United States in 1933.

a   Study the figures carefully and describe the apparent relationship, if any, between the size of newspaper circulation areas and population density. How does this relate to Isard's Theory?

b   Describe the data *you* would need to collect, and the procedures *you* would use, to test the relationship more rigorously.

c   Describe and explain the correspondence (or lack of it) between a few state boundaries and the boundaries of the newspaper circulation fields.

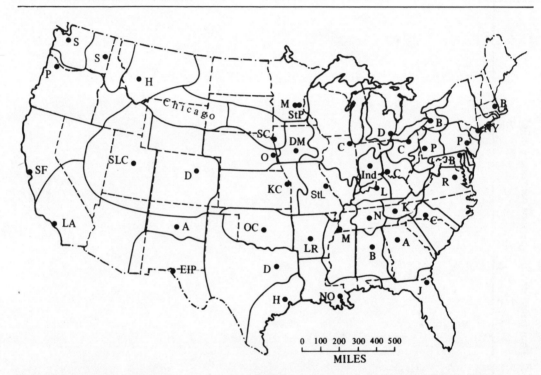

Figure 2.25   The circulation areas of metropolitan newspapers in the United States, 1933
Reprinted with permission from R. E. Park and C. Newcomb (1933) in *The Metropolitan Community* edited by R. D. McKenzie, published by the McGraw-Hill Book Company, figure 6.

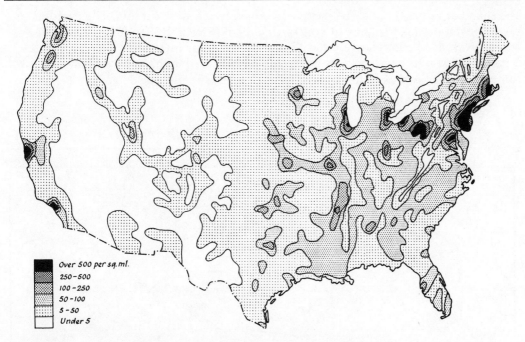

Figure 2.26   The pattern of population density in the United States
Reprinted with permission from *A Social Geography of the United States* by J. Wreford Watson
(1979) published by Longman, London and New York, figure 1.2.

THE SHAPE OF TERRITORIES: THE CASE OF ENGLISH PARISHES

As we have seen, the ideal shape for packed territories in theory is the hexagon. This thesis can be put to the test by measuring the number of border contacts between one territory and adjacent ones. For instance, in Figure 2.27, which displays some parishes west of Norwich, the parish of Sporle, reference number 25, contacts seven other parishes around its border.

1   Calculate the contact numbers for all parishes shown in Figure 2.27. Calculate the mean number of contacts per parish. Plot the results as a bar graph (Figure 2.29); by way of example, the data for parishes east of Swindon (Figure 2.28) has been plotted for you.

2   What do the results suggest?

The shape of territories may be partly fashioned by physical factors. The parishes east of Swindon lie in an area of contrasting relief (Figure 2.30), whereas the parishes west of Norwich lie in relatively flat terrain. It would be interesting therefore to see if parish shape in the two areas is different. We may define shape by an index of compactness defined as follows

$$\text{Index of compactness} = \frac{R_i}{R_c}$$

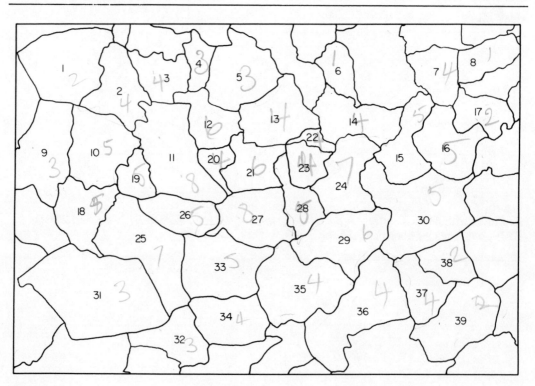

Figure 2.27   Parishes west of Norwich (list of names on p. 120)

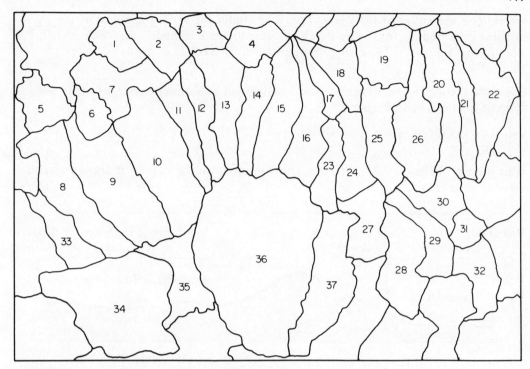

Figure 2.28   Parishes east of Swindon (list of names on p. 120)

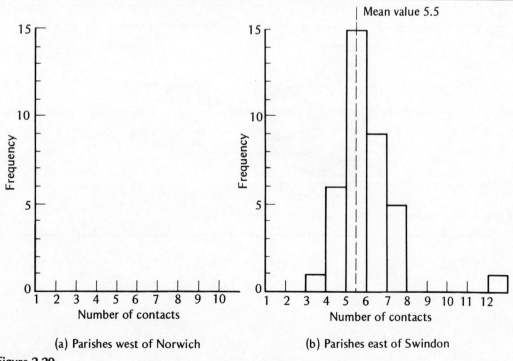

(a) Parishes west of Norwich   (b) Parishes east of Swindon

Figure 2.29

118

where R_i is the radius of the largest inscribed circle, and R_c is the radius of the smallest inscribed circle (Figure 2.31). A circle yields a compactness index of 1.0, a hexagon a value of 0.75; a long thin parish would have a compactness index of about 0.1.

3 Calculate the index of compactness for each parish and present your results in the form of histograms on Figure 2.32.

4 Calculate the mean values for each area and draw them on your histograms as shown in Figure 2.33.

5 With the aid of Figure 2.33, interpret and comment on the degree of dispersion and skew of your results.

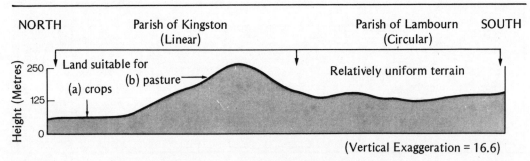

Figure 2.30   A sketch section across two parishes in the Berkshire Downs

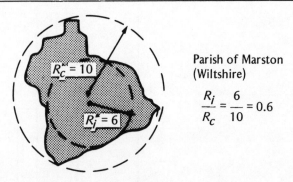

Figure 2.31   Deriving the index of compactness

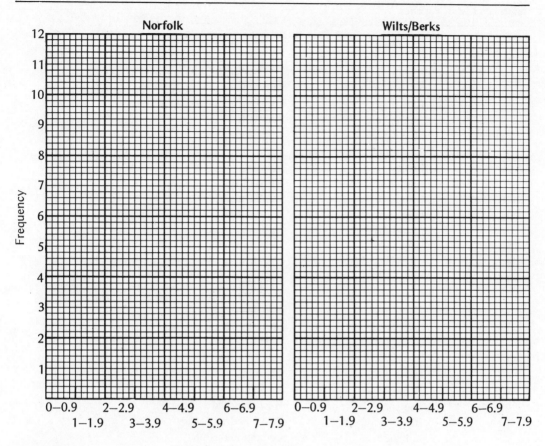

Figure 2.32   Index of compactness (class intervals)

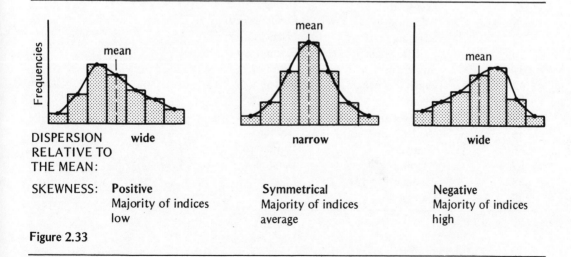

Figure 2.33

Reference List of Parishes

Reference No.	Norfolk	Berks/Wilts
1	Gt. Massingham	Watchfield
2	Rougham	Longcot
3	Weasenham	Fernham
4	Wellingham	Baulking
5	Tittleshall	Marston
6	Brisley	Bourton
7	Billingford	Shrivenham
8	Foxley	Wanborough
9	West Acre	Bishopstone
10	Castle Acre	Ashbury
11	Lexham	Knighton
12	Litcham	Woolstone
13	Mileham	Uffington
14	Beetley	Kingston
15	Hoe	Sparsholt
16	Swanton Morley	Childrey
17	Bylaugh	West Challow
18	South Acre	East Challow
19	Newton	Grove
20	Kempstone	Ardington
21	Beeston	West Hendred
22	Billering	East Hendred
23	Longham	Letcombe Bassett
24	Gressenhall	Letcombe
25	Sporle	Wantage
26	Dunham	Lockinge
27	Frensham	Fawley
28	Wendling	Chaddleworth
29	Scarning	Brightwalton
30	East Dereham	Farnborough
31	Swaffham	Catmore
32	North Pickenham	Peasemore
33	Necton	Liddington
34	Holme Hale	Aldbourne
35	Bradenham	Baydon
36	Shipdam	Lambourn
37	Whinborough	Garston
38	Yaxham	
39	Garrestone	

Further Reading

Inside the City, J. Everson and B. FitzGerald, Longman (1972), Chapters 5 and 8.

Human Geography: Theories and Their Applications, M. G. Bradford and W. A. Kent, Oxford University Press (1977), Chapter 8.

Pattern and Process in Human Geography, V. Tidswell, University Tutorial Press (1976), Chapter 12.

The North American City, M. H. Yeates and B. J. Garner, Harper and Row (1976), Chapter 3.

Analytical Human Geography, P. Ambrose, Longman (1972), Chapter 4.

Techniques in Human Geography, P. Toyne and P. Newby, Macmillan (1972), Chapter 2.

CHAPTER THREE
URBAN STRUCTURE

Urban Structure

Theories of urban structure have their roots in the Chicago school of urban geography led by R. E. Park, E. W. Burgess, and R. D. McKenzie in the 1910s and 1920s. Park, a journalist who had visited many of the world's cities, developed a theory of city structure based on an analogy between plant communities and human communities. He translated a number of ecological processes into human terms. First and foremost of these was the concept of competition. Within a city man competes for the most desirable site to live and the most desirable site for business activities. Desirable sites command higher land values than undesirable ones. So competitive activity, through the land price mechanism, sorts like people into distinct areas, such as the central business district, areas of trade, and residential areas, according to how much rent they can afford.

Another process translated by Park was dominance. In the same way that oak trees are the dominant element in a fully developed oak woodland, their height and leaves acting to influence the plant species found on the woodland floor, the central business district is the dominant element in an urban area. Competition between businesses to locate in the highly accessible central business district gives rise to high land values there which in turn influence the location of other elements in the city. Within local areas, other activities are dominant — industry in industrial areas, high income residency in high status areas, and so on.

A third ecological process interpreted by Park in an urban setting was invasion and succession. Plants which invade a fresh soil surface alter the environment in their vicinity and create conditions more favourable to less tolerant plants which are then able to invade the area and oust the pioneer species. The process of invasion may continue through a succession of plant species until a climax association is established in which no further change in the dominant plant species takes place. The process of invasion and succession in human communities was seen by Park in the invasion of residential areas by commercial and business concerns, in the invasion of higher status residential areas by lower income groups, and, for Chicago, the invasion of an area of one ethnic group by people from another ethnic group.

Also derived from the ecological analogy was the concept of gradient. In an organism, the brain develops in the most active region and other organs form in a definite order in particular places. As a parallel to this in the growth of cities, the gradient of land values declines outwards from a peak at the city centre, the most active region in the city. Owing to the sifting and sorting effect of land values, many other social phenomena — income levels, rates of social disorder, and crime — assume the same gradient.

URBAN ACTIVITY AND LAND VALUES

The land-use pattern of a city reflects both the contrasting site requirements of urban activities and the competition for these sites through the price system. For example, retailing is obviously dependent upon contact with potential customers and seeks sites within areas of heavy pedestrian traffic. Accessibility measures the ease with which such contact can be made, the most accessible locations corresponding to the focal points of the road and rail routes. The location of retail activity can be thought of as an economic commodity subject to the forces of supply and demand. Given that accessible locations near the city centre are scarce and also highly desirable, the market price of such sites tends to be far higher than in surrounding city areas. Other accessible areas of land which fetch high market prices are found along major radial road routes or at suburban shopping centres (for instance, Figure 3.1).

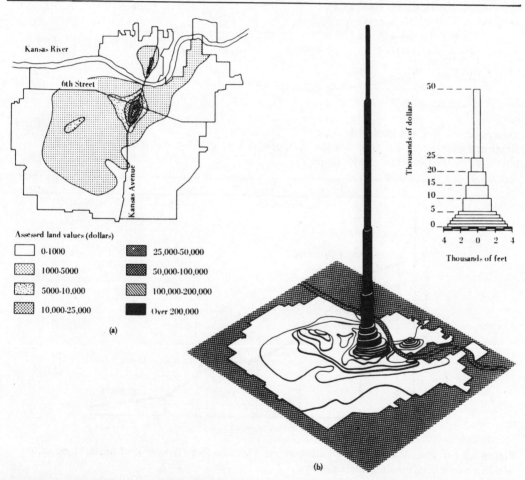

Figure 3.1 The distribution of land values in Topeka, Kansas

Reprinted with permission from *The Distribution of Land Values in Topeka, Kansas*, by D. S. Knos (1962) published by The University of Kansas Institute for Social and Environmental Studies, figures 1 and 2.

The pattern of urban land use therefore reflects the pattern of urban land prices, the more accessible, desirable sites being used by activities which can afford the high prices. Figure 3.2 illustrates bid price (the price which could be offered for land in auction) gradients of three urban activities, based on their ability to benefit from a city centre location.

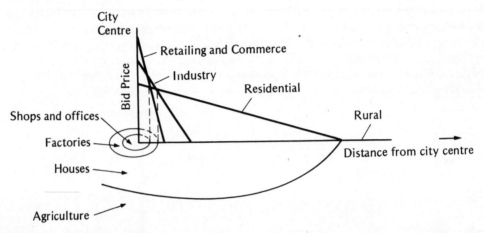

Figure 3.2 Bid-rent gradients
Reprinted with permission from *Section II, Spatial Analysis — Area Patterns,* Units 15–17, published by the Open University, figure 11.

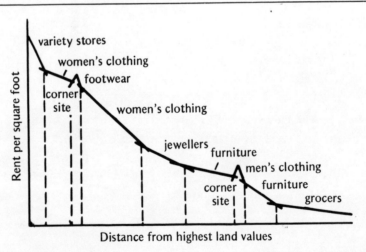

Figure 3.3 Hypothetical rent gradient in one direction from the point of highest land values within a major unplanned shopping centre
Reprinted with permission from *Geography and Retailing* by P. Scott (1970) published by Hutchinson, figure 1.

1 With reference to Figure 3.3, suggest why different types of retailer locate at various distances from the point of peak land value.

In addition to being segregated by market forces, similar urban activities are pulled together by complementary linkages. For example, shops may increase their turnover when clustered together since more potential customers may be attracted where a greater variety of merchandise is available. Similarly, a clustering of offices might be indicative of a profitable location and therefore attract more office development wishing to capitalize on what is apparently a good site for offices.

POPULATION DENSITY VARIATIONS

In the previous exercise we saw that land values decline from a city centre. Related to this is a decrease in intensity of land use and population density.

1 With reference to Figure 3.4, construct a choropleth map to show the variation of population density in Norwich.

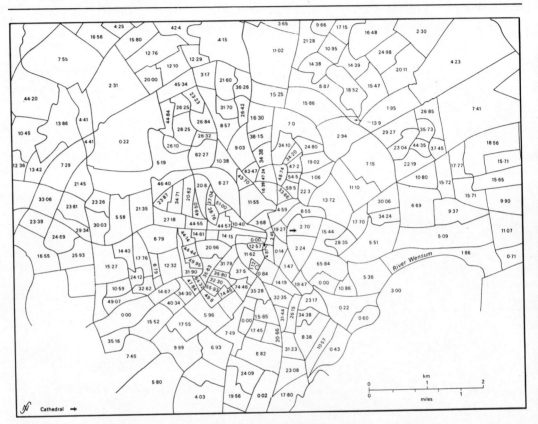

Figure 3.4 Population density variations in Norwich. Units are persons per acre
Reprinted with permission from Norwich City Planning Department.

2 Is there a uniform decline in population density? Can you suggest reasons for an uneven decline in density?

Colin Clark (1951) observed that, regardless of time or place, city population density declines exponentially away from the city centre. Moreover, he found that as cities grow they experience a decline in population density gradients. At first, most people live in the central zone and the number of people per unit area declines rapidly away from the city centre. With time people tend to move out of the central city area, suburbs grow, and the number of people per unit area, though usually still largest in the central area, declines far less rapidly as the rural population density is approached (Figure 3.5).

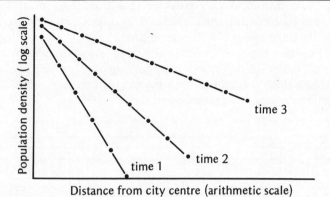

Figure 3.5   Negative exponential population decline from city centre

3   Using data from Table 3.1, construct a graph on Figure 3.6 to show the population
density gradient for Greater London. How far does the pattern conform with Clark's
observations? A negative exponential relationship would appear as a straight line on
the graph.

Table 3.1

Distance from the City of London (km)	Average population density (persons per hectare)
0	15.5
2	92.5
4	124.2
6	109.1
8	111.6
10	101.3
12	80.0
14	66.3
16	47.5
18	42.6
20	40.1
22	37.2
24	32.5
26	27.1
28	19.3

In attempting to explain the pattern of population density, W. Alonso has shown
that the bid-rent functions are steeper for the less wealthy of any pair of households
with identical tastes. Therefore one would expect the more affluent city dwellers
who can afford high commuting costs and who live at low residential densities
(few houses per unit area) to be found on the edge of the city, whereas the less

affluent would be found in inner city areas living at high residential densities; this situation would lead to an exponential decrease in population density away from the city centre. Also, Richard Muth (1960) has argued that if the price-distance function is assumed to be negative exponential then the population density must itself decline in a similar manner. Brian Berry provides supporting evidence from Chicago indicating that residential land values (measured in terms of front footage) decline exponentially beyond the inner city zone of redevelopment.

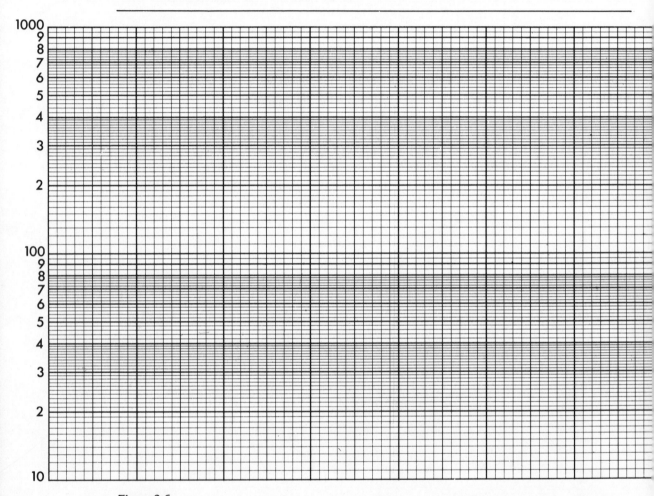

Figure 3.6

An alternative method of illustrating changes in population density gradient is to construct a density-distance scattergraph and plot the regression line. A regression line is an accurate best-fit line which describes a relationship between two variables (such as population density and distance) mathematically. In this context it is particularly useful because the change in the scatter of points may be difficult to detect by eye. The rate of decline of population density with distance is measured by the slope of the line which is denoted by the letter b (this can be thought of as tangent of the angle θ in Figure 3.7). The larger is b, the steeper is the city population density gradient. Therefore, from what we have already discussed, we should expect b to be greater in the early stages of city growth than in the later stages. The letter a describes the position of the line on the graph by giving the point at which it cuts the y axis.

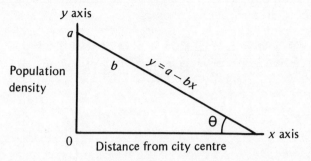

Figure 3.7 Elements of a regression line (inverse relation)
Note: *b* will only be equal to the tangent of the angle if the axes are plotted on the same scale. For example 1 cm = 1 person per hectare = 1 km; clearly this would be an inconvenient scale to use.

A Worked Example: Norwich

Using 1961 ward data for Norwich (Table 3.2) the following procedure may be applied.

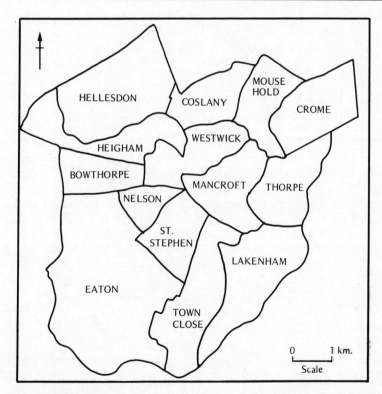

Figure 3.8 Norwich wards

Table 3.2 Data for calculating regression line for Norwich ward, population density data, 1961

1 Ward	2 Distance from city centre, x (km)	3 Population density, y (persons per hectare)	4 $(x - \bar{x})$	5 $(y - \bar{y})$	6 $(x - \bar{x})^2$	7 $(y - \bar{y})^2$	8 $(x - \bar{x})(y - \bar{y})$
Bowthorpe	2.3	47.69	0.46	5.05	0.2116	25.4016	2.3184
Coslany	1.6	66.96	−0.24	24.31	0.0576	590.9761	−5.8344
Crome	2.8	39.50	0.96	−3.15	0.9216	9.9225	−3.0240
Eaton	3.1	21.00	1.26	−21.65	1.5876	468.7225	−27.2790
Heigham	2.5	16.80	0.66	−25.85	0.4356	668.2225	−17.0610
Hellesdon	2.7	21.25	0.86	−21.40	0.7396	457.9600	−18.4040
Lakenham	1.8	33.61	−0.04	−9.04	0.0016	81.7216	0.3616
Mancroft	0.0	31.63	−1.84	−11.02	3.3856	121.4404	20.2768
Mousehold	1.7	49.17	−0.14	6.52	0.0196	42.5104	−0.9128
Nelson	1.4	78.33	−0.44	35.68	0.1936	1273.0624	−15.6992
St. Stephen	1.3	64.74	−0.54	22.09	0.2916	487.9681	−11.9286
Thorpe	1.5	32.37	−0.34	−10.28	0.1156	105.6784	3.4952
Town Close	2.3	36.82	0.46	−5.83	0.2116	33.9889	−2.6818
Westwick	0.8	57.32	−1.04	14.67	1.0816	215.2568	−15.2568
Totals	25.8	597.19			9.2544	4582.7843	−91.8966
Means	$\bar{x} = 1.84$	$\bar{y} = 42.65$					

i List the wards, their populations and distance of them away from the city centre in columns 1, 2 and 3.

ii Calculate the mean distance of the wards ($\bar{x}$) from the city centre by adding the distances of the wards (x) from the centre and dividing by the total number of wards (n); hence $\bar{x} = \dfrac{\Sigma x}{n}$.

iii Complete column 4 by subtracting the mean distance of wards from the city centre ($\bar{x}$) from the distance of each ward from the centre (x).

iv Calculate the mean population ($\bar{y}$) where $\bar{y} = \dfrac{\Sigma y}{n}$.

v Complete column 5 by subtracting the mean population ($\bar{y}$) from each ward population (y).

vi Square each of the values in column 4 and place in column 6, remembering that a negative value when squared becomes positive.

vii Square each of the values in column 5 and place in column 7.

viii Multiply each of the values in column 4 by the values in corresponding rows of column 5 and place in column 8. (*Note*: Some of these may be negative values)

ix Sum columns 6, 7 and 8. In statistical notation these are expressed as $\Sigma(x - \bar{x})^2$; $\Sigma(y - \bar{y})^2$ and $\Sigma(x - \bar{x})(y - \bar{y})$ respectively.

x The slope of the regression line will then be given by the formula

$$b = \frac{\Sigma(x - \bar{x})(y - \bar{y})}{\Sigma(x - \bar{x})^2} = \frac{\text{sum of column 8}}{\text{sum of column 6}}$$

$$b = \frac{-91.8966}{9.2544} = -9.93 \text{ persons/ha/km}$$

which means that for every kilometre increase away from the city centre, the population density decreases by 9.93 persons per hectare.

xi To draw the regression line the value of population density at the city centre must be found (that is a — see Figure 3.7). This is given by the equation

$a = \bar{y} - b\bar{x}$
   $= 42.65 - (-9.93 \times 1.84)$
   $= 42.65 + 18.2712$
   $= 60.92$ persons per hectare.

In other words when $x = 0$
$$y = 60.92.$$

xii Substitute the value a in the equation of the regression line

$y = a - bx$

which in our example is

$y = 60.92 - 9.93x$

xiii Having calculated the values of a and b, we need to fix another point on the graph in order to draw the regression line.

For example, when x is 3.0 km
$$y = 60.92 - (9.93 \times 3)$$
$$= 31.13 \text{ persons/hectare.}$$

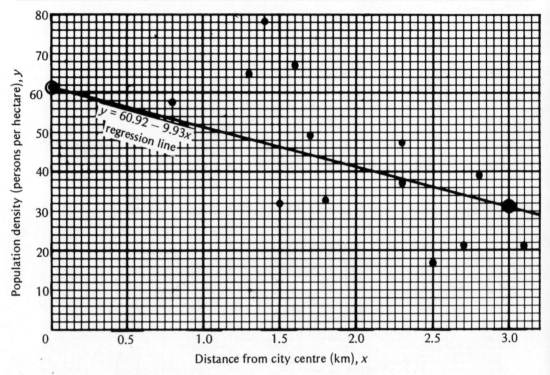

Figure 3.9   Population density gradient for Norwich, 1961

4 Using 1971 data from Table 3.3, calculate and draw the regression line on Figure 3.10. The regression line for 1961 has been drawn for easy comparison. Comment on your results.

Table 3.3

x y

Ward	Distance from centre (km)	Population density (persons/ha)
Bowthorpe	2.3	45.39
Coslany	1.6	65.24
Crome	2.8	48.47
Eaton	3.1	25.17
Heigham	2.5	44.32
Hellesdon	2.7	24.61
Lakenham	1.8	35.01
Mancroft	0.0	20.25
Mousehold	1.7	45.93
Nelson	1.4	77.53
St. Stephen	1.3	60.09
Thorpe	1.5	32.12
Town Close	2.3	31.51
Westwick	0.8	28.71

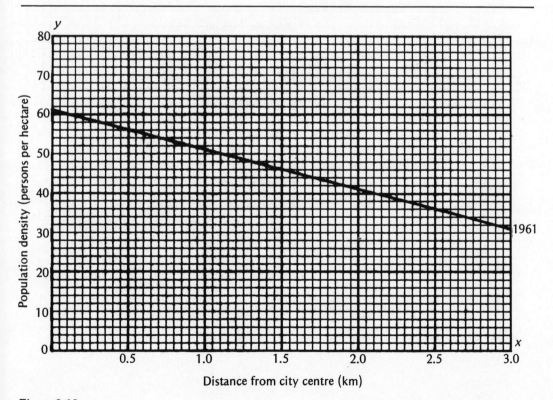

Figure 3.10

THE VARIATION OF BUILDING AGE IN WARE

Given the historical development of most settlements, building age usually varies within towns and building-age gradients can be identified. The town of Ware in Hertfordshire situated on the River Lea and London-to-Cambridge road, originated as an important malting centre in the sixteenth century. Since that time, economic forces have been too weak to overcome the inertia of the capital invested in the buildings and preservation orders make it increasingly difficult to replace them with modern ones at higher densities.

1a   Examine Table 3.4* which provides data on the percentage of each hectare occupied by building units of different age groups. In conjunction with Figure 3.11 *either*, by working in groups, construct a series of maps (choropleth) to show the variation in building age in a single map for each age group.

b   *Or* construct a choropleth map to show the dominant age group in each square. To what extent do these results conform to theories of urban structure?

c   *Or* illustrate this feature by a composite graph (Figure 3.12), using the data shown in Table 3.5.

Table 3.5

Distance from centre (hectometres)	Number of building units						
	15th century	16–17th century	18th century	Victorian	Edwardian	Inter-war	Post-war
0–0.25	4	41	16	20	4	4	11
1.25	2	98	7	11	10	2	92
2.25	2	15	4	90	30	0	90
3.25	0	10	4	123	60	64	77
4.25			2	182	62	114	144
5.25			2	84	11	161	260
6.25			5	48	30	145	380
7.25					23	74	444
8.25					12	25	192
9.25					2	8	201
10.25					2	8	47
11.25						7	19
12.25						6	

*
The data for this exercise were kindly provided by Jonathan Ratter.

2 The percentage of hectare domination by four age categories is in Table 3.6. What relationship does this data reveal? Can you suggest reasons for the variety in building density?

Table 3.6   (Figures as percent for each age group)

	18th and earlier	Victorian/Edwardian	Interwar	Postwar
10 per ha	50	61	82	50
20 per ha	40	27	18	35
30 per ha	10	6	—	13
40 per ha	—	6	—	2
	100	100	100	100

Table 3.4

Grid number	15th century and earlier	16th and 17th century	18th century	Victorian	Edwardian	Interwar	Postwar
A 1						34	66
2						100	
3							60
4				95		5	
5							
6						90	10
7						97	3
8						44	56
9							100
10							100
11							100
12							100
13					62	7	31
14					11		89
15						70	30
16						62	38
17						100	
18							100
19							100
20							100
21							

Grid number	15th century and earlier	16th and 17th century	18th century	Victorian	Edwardian	Interwar	Postwar
B 1							
2							
3							
4						66	34
5			20		22	58	
6						95	5
7				10		90	
8						40	60
9							100
10					90	10	
11			34		66		
12					18		82
13					91		9
14					32	4	64
15							100
16				11	27		72
17							100
18							100
19							100
20							100
21							100

Table 3.4   continued

Grid number	15th century and earlier	16th and 17th century	18th century	Victorian	Edwardian	Interwar	Postwar	Grid number	15th century and earlier	16th and 17th century	18th century	Victorian	Edwardian	Interwar	Postwar
C 1								D 1							
2								2							
3								3							100
4						100		4						100	
5						100		5						100	
6						100		6				37		25	38
7						100		7						50	50
8						66	34	8			2		48	16	32
9				66	20		14	9				14	86		
10	7			14	21		58	10				60		35	5
11				25			75	11	25				18		43
12				14	16		70	12					50		50
13			100					13					33	50	16
14			100					14					70		30
15				90			10	15				23	14	12	51
16				8		92		16						55	45
17							100	17							100
18							100	18							100
19							100	19						4	96
20							100	20							100
21							100	21							

Table 3.4   continued

Grid number	15th century and earlier	16th and 17th century	18th century	Victorian	Edwardian	Interwar	Postwar
E 1							
2							
3							100
4						50	50
5						50	50
6							100
7				100			
8					100		
9	22			22	56		
10	62	6	6	13			13
11	55	4		16			25
12				25	12		63
13					48	5	47
14			5	15	5	25	50
15				18			82
16			30		13	34	23
17						10	90
18						55	45
19				25			75
20							
21							

Grid number	15th century and earlier	16th and 17th century	18th century	Victorian	Edwardian	Interwar	Postwar
F 1							
2							
3							
4							
5						100	
6					33	66	
7						100	
8			16	16	52		16
9		92			8		
10		16			84		
11	11	33	11		23	11	11
12			56		33	17	
13					85	7	7
14						7	93
15				25	10	5	60
16				15	70	8	7
17				8	60	8	24
18				50	7	4	39
19				65		10	25
20							
21							

Table 3.4   continued

Grid number	15th century and earlier	16th and 17th century	18th century	Victorian	Edwardian	Interwar	Postwar	Grid number	15th century and earlier	16th and 17th century	18th century	Victorian	Edwardian	Interwar	Postwar
G1								H1							
2								2							
3								3							
4							100	4							
5					14	86		5							100
6			100					6							
7		16		52			32	7				100			
8							100	8							
9	22	23					55	9							
10		20	45	15	20			10				100			
11		60	10		16	7	7	11				100			
12		80	12	2	2	2	2	12	7	60	7	7		7	12
13	3	15	3		79			13		44		4	8	4	40
14					32	12	56	14				100			
15					45	15	40	15					75	25	
16					10	90		16					100		
17				14	50	36		17					100		
18				40		10	50	18				31	11	25	33
19							100	19				10			90
20						40	60	20						10	90
21								21						34	66

Table 3.4   continued

Grid number	15th century and earlier	16th and 17th century	18th century	Victorian	Edwardian	Interwar	Postwar	Grid number	15th century and earlier	16th and 17th century	18th century	Victorian	Edwardian	Interwar	Postwar
I 1								J 1							
2								2							
3								3							
4								4							
5								5							
6								6							
7								7							
8								8							
9							100	9							
10							100	10							
11							100	11							
12			50				50	12				38		50	12
13		16	5	10	5		64	13				7	21	42	30
14		9	30	30	30			14		30		20	25		25
15					33	33	34	15					100		
16			21	35	21	23		16				30			60
17		14	6		6		74	17							100
18								18							
19						90	10	19							
20							100	20							
21								21							

Table 3.4   continued

Grid number	15th century and earlier	16th and 17th century	18th century	Victorian	Edwardian	Interwar	Postwar	Grid number	15th century and earlier	16th and 17th century	18th century	Victorian	Edwardian	Interwar	Postwar
K 1								L 1							
2								2							
3								3							
4								4							
5								5							
6								6						15	85
7				12	88			7						80	20
8					60	56		8							
9				10	45	45		9					9		42
10					78	22		10							96
11							100	11							100
12	4	8		4			84	12				3		15	82
13					16	84		13							100
14				80	20			14						66	33
15			34	66				15					50		50
16							100	16							
17						100		17							
18								18							
19								19							
20								20							
21								21							

Table 3.4   continued

Grid number	15th century and earlier	16th and 17th century	18th century	Victorian	Edwardian	Interwar	Postwar
M1							
2							
3							
4							
5							100
6							100
7							
8							100
9					84		16
10					30	10	60
11						20	80
12						33	66
13						40	60
14					100		
15							
16							
17							
18							
19							
20							
21							

Grid number	15th century and earlier	16th and 17th century	18th century	Victorian	Edwardian	Interwar	Postwar
N1							
2							
3							
4							100
5							
6							
7							
8							
9							100
10							100
11							100
12							
13							
14							
15							
16							
17							
18							
19							
20							
21							

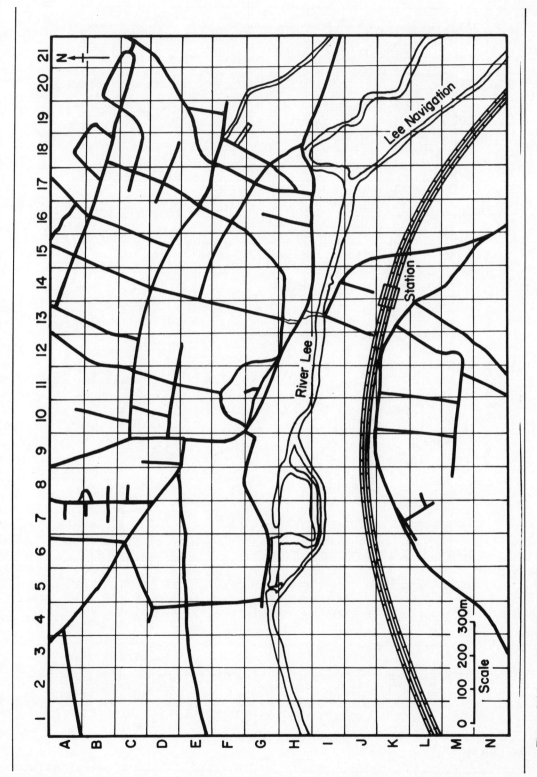

Figure 3.11   Base map of Ware

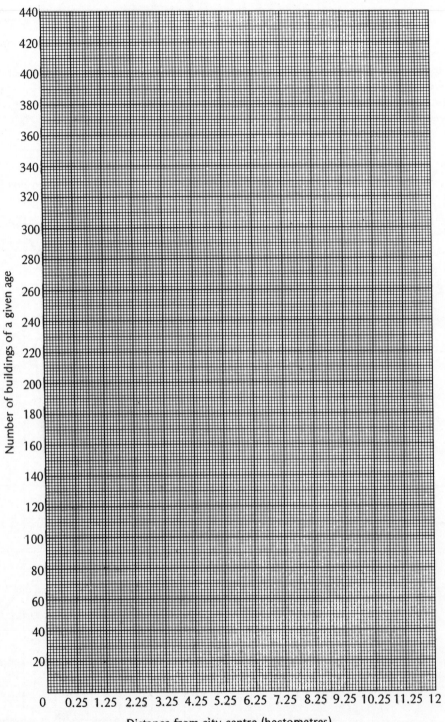

Figure 3.12

Land-Use Models

As early as 1923, R. D. McKenzie suggested that Columbus, Ohio, could be divided into concentric circles which represented the form of the city's growth. In 1925, E. W. Burgess, combining the ecological analogy with McKenzie's rings, put forward his now classic concentric-ring model of the growth and structure of Chicago. He envisaged the process of invasion and succession operating within the city. Higher status groups migrated from the old inner city to the better outer city areas, and so started an outwards movement of people away from the city centre. Residential zones grew by invading and taking over the inner part of the next outer ring. W. Alonso argued that the centrifugal movement of people was a result of the willingness to forgo the advantages of living near the city centre and instead live in an area of low-density housing.

Burgess's model consists of five concentric rings (Figure 3.13). The first and innermost ring is the central business district, known to Chicago locals as the Loop. The second ring, a zone in transition, has been invaded by commerce and businesses which lower its residential desirability; it is characterized by a mixture of land use — industry, commerce, business, high-density residencies of low-income groups, and residences of "undesirables" such as prostitutes; and being an area of cheap accommodation, it is a zone invaded by newly arriving ethnic groups. The third ring is a zone of small, inexpensive frame houses of working men who had originally lived in the second ring but, having become more prosperous, moved out; this ring thus grew by the process of natural invasion and succession. The fourth ring is a zone containing the better dwellings of middle-class white-collar workers and professional people; it thus continues the rise in status with distance from the city centre. The outermost ring, the zone of commuters, is predominantly an affluent suburban area.

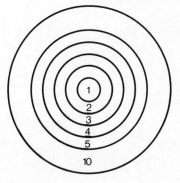

1	Central business district
2	Wholesale, light manufacturing
3	Low-class residential
4	Medium-class residential
5	High-class residential
10	Commuters' zone

Figure 3.13   Burgess's concentric-ring model of urban structure
Reproduced by permission from the *Annals* of the Association of American Geographers, Volume 54, 1964, A. R. Pred.

Chicago is not the only city whose structure may be interpreted using Burgess's concentric-ring model. Some concentric rings have been recognized in, for example, Philadelphia in the United States and Calgary in Canada.

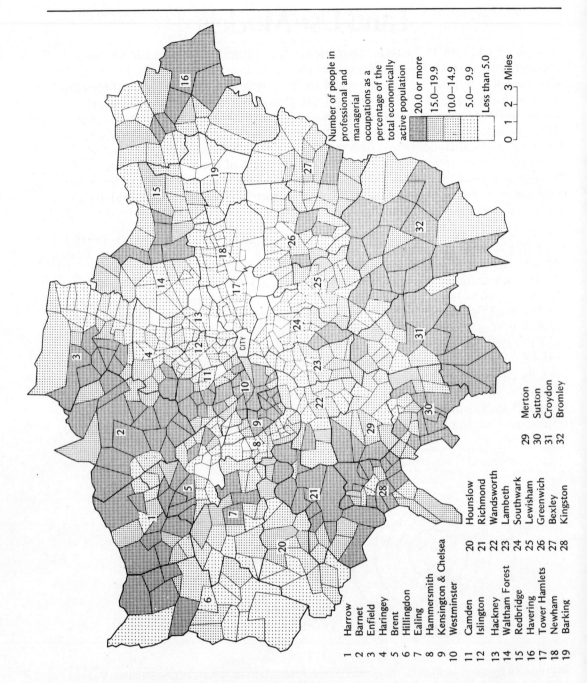

Figure 3.14 The distribution of the professional and managerial socio-economic group in London Boroughs
Reprinted with permission from *Social Atlas of London* by J. Shepherd, J. Westaway and T. Lee (1974) published by Oxford University Press, map 5.1.

1   Examine Figure 3.14 and describe and attempt to explain the distribution of the professional and managerial socio-economic group.

The Burgess model did not pass unchallenged. In 1938, M. R. Davie took a detailed look at twenty cities in the United States and concluded that lines of communication, together with the industry they tend to attract, are the dominant factor in explaining residential patterns or urban areas. Low-grade housing is associated with the industrial areas near lines of water and rail transport. A year later, a report made by Homer Hoyt, whose study was based on rent data collected from a large number of cities, put forward a new model of urban structure. Focusing on the spread of high-status residential areas, which he thought determined the disposition of other areas in the city, Hoyt found an outwards movement originating close to the city centre where the higher income population work, and spreading along radial transport routes which have good commuting facilities, along high ground which is free from flooding, along waterfronts not already occupied by industry, and into open country. The result is that high-status sectors develop in the city which pull the growth of other city areas in the same direction (Figure 3.15). Though Hoyt emphasized a directional element in city expansion which leads to sectors of growth and segregation, he also recognized the outwards, centrifugal movement which is the chief process operative in Burgess's model — he saw the building of new residencies on the periphery of the city as the driving force which draws people from the older houses and causes all groups to move up a step, leaving the oldest and cheapest houses vacant for the lowest income families and undesirables. Hoyt differed from Burgess in suggesting that the invasion process varied from sector to sector, thus from one kind of social area to another.

The sectoral arrangement of land use in cities has been recognized in Chicago (obviously using a different interpretation to Burgess) and other cities including Belfast.

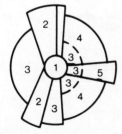

1   Central business district
2   Wholesale, light manufacturing
3   Low-class residential
4   Medium-class residential
5   High-class residential

Figure 3.15   Hoyt's sectoral model of urban structure
Reproduced by permission from the *Annals* of the Association of American Geographers, Volume 54, 1964, A. R. Pred.

150

2a Examine Figures 3.16 and 3.17, and comment on the pattern of land use with reference to the Burgess and Hoyt models.

b To what extent is the presence and course of the River Wear likely to have influenced the pattern of land use?

c How does the presence of local authority housing affect the applicability of such models?

d According to Muth (1965), residential land values are depressed near industry because the environmental disadvantages more than offset the advantages of accessibility. In contrast, local shops, parks, and beaches, etc., appear to have a favourable effect on land values. How true is this of Sunderland?

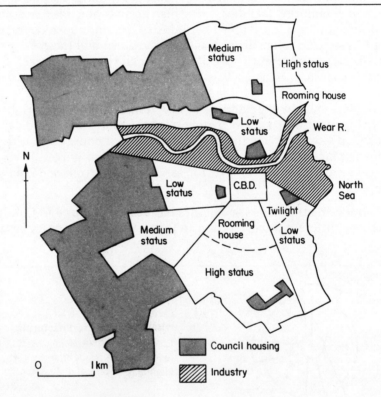

Figure 3.16 Social areas in Sunderland
Based on work of B. T. Robson (1969) *Urban Analysis* published by Cambridge University Press. Reprinted in a slightly modified version with permission from *Urban Residential Patterns* by R. J. Johnston (1971) published by Bell & Hyman, figure IV.6.

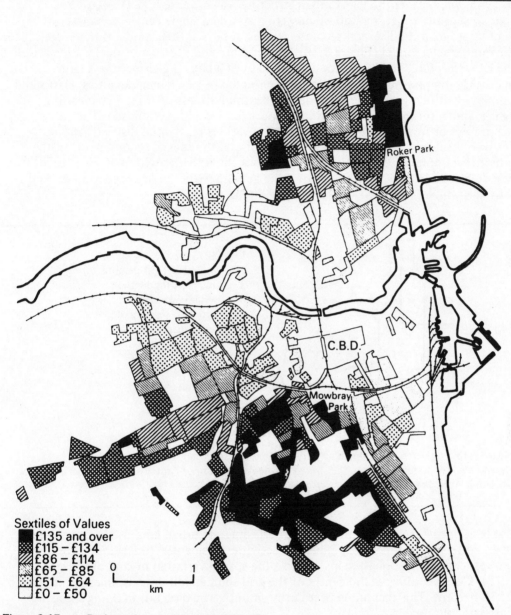

Sextiles of Values

- £135 and over
- £115 – £134
- £86 – £114
- £65 – £85
- £51 – £64
- £0 – £50

Figure 3.17   Rating values in Sunderland, 1963
Reprinted with permission from "An ecological analysis of the evolution of residential areas in Sunderland" by B. T. Robson (1966), *Urban Studies*, 3, figure 3.

In 1945, yet another model of urban structure was proposed: C. D. Harris and E. L. Ullman suggested that, instead of clustering around a single centre, urban activities tend to group around several separate centres or nuclei; their model is therefore called the multi-nuclei model (Figure 3.18). The number and position of nuclei depends on the size, structure, and historical development of a city. Big cities have a larger number and more specialized centres than small cities. In American cities, Harris and Ullman identified five nuclear districts: a central business district; a wholesaling and light-manufacturing area near main transport routes; a heavy industry district near the present or former outer edge of the city; various residential districts; and outer dormitory suburbs. Each district grows because of the special requirements of the activities it contains. Clearly it is advantageous for activities with like needs to locate near to one another. The presence of some activities will tend to repel others. And activities which can afford similar rents will tend to cluster.

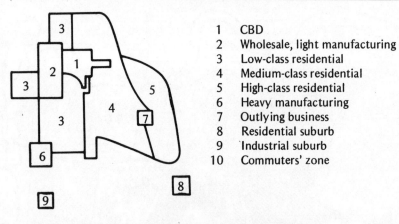

1	CBD
2	Wholesale, light manufacturing
3	Low-class residential
4	Medium-class residential
5	High-class residential
6	Heavy manufacturing
7	Outlying business
8	Residential suburb
9	Industrial suburb
10	Commuters' zone

Figure 3.18 Harris and Ullman's multi-nuclei model of urban structure
Reproduced by permission from the *Annals* of the Association of American Geographers, Volume 54, 1964, A. R. Pred.

Table 3.7 provides information on the percentage of urban land use in Dorchester devoted to service functions, industrial activities, residential functions.* Working in groups, construct isopleth maps to show the spatial variation in land use (Figure 3.19). Plot the values at the centre of the grid squares and draw isolines at intervals of 25 percent. This technique necessarily smooths the data but in doing so distinct patterns are revealed.

Discuss the patterns of land use revealed.

*
The data for this exercise were kindly provided by Neville Grenyer.

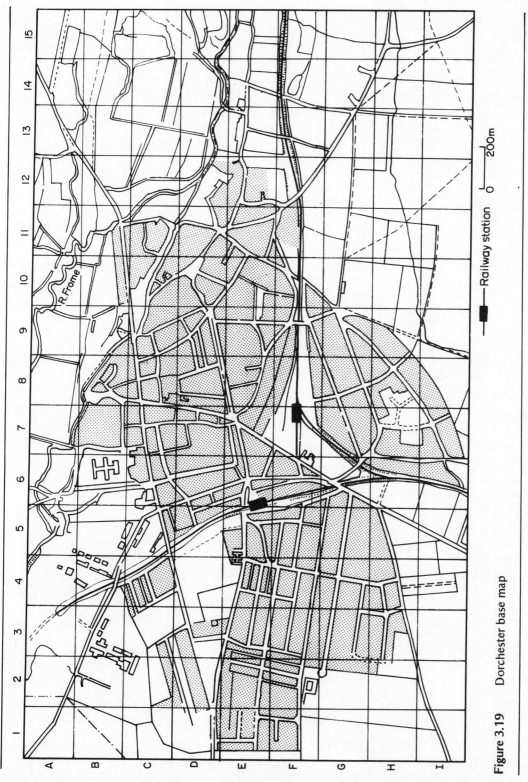

R. Frome

Railway station

0 200m

Figure 3.19   Dorchester base map

Table 3.7

Land use in Dorchester (percent)

Grid square	Services	Industry	Residential	Other	Grid square	Services	Industry	Residential	Other	Grid square	Services	Industry	Residential	Other
A1	0	0	0	100	D1	2	0	98	0	G1	0	0	0	100
A2	0	0	0	100	D2	0	0	100	0	G2	0	0	25	75
A3	0	0	0	100	D3	14	0	86	0	G3	0	0	100	0
A4	100	0	0	0	D4	10	0	0	100	G4	5	0	55	0
A5	2	57	41	0	D5	0	0	100	0	G5	0	0	100	0
A6	6	13	63	18	D6	0	0	97	3	G6	22	1	77	0
A7	0	0	0	100	D7	9	0	14	77	G7	0	0	96	4
A8	0	0	0	100	D8	1	14	60	25	G8	0	0	98	2
A9	0	0	0	100	D9	0	0	100	0	G9	0	0	100	0
A10	0	0	0	100	D10	1	0	93	6	G10	0	0	0	100
A11	0	0	0	100	D11	3	13	84	0	G11	0	0	0	100
A12	0	0	0	100	D12	9	0	41	50	G12	0	0	0	100
A13	0	0	0	100	D13	0	0	0	100	G13	0	0	100	0
A14	0	0	0	100	D14	0	0	0	100	G14	0	0	0	100
A15	0	0	0	100	D15	0	0	0	100	G15	0	0	0	100
B1	0	0	0	100	E1	3	0	97	0	H1	0	0	0	100
B2	100	0	0	0	E2	0	0	100	0	H2	0	2	10	90
B3	100	0	0	0	E3	1	0	99	0	H3	0	0	50	50
B4	100	0	0	0	E4	9	1	85	5	H4	0	0	50	50
B5	31	0	65	1	E5	9	1	85	5	H5	0	0	20	80
B6	60	0	40	0	E6	0	3	97	0	H6	7	0	20	80
B7	7	5	79	11	E7	2	5	93	0	H7	0	0	100	0
B8	3	0	91	6	E8	4	0	96	0	H8	2	0	9	7
B9	0	18	24	58	E9	25	0	95	0	H9	0	0	100	50
B10	0	33	0	67	E10	24	0	76	0	H10	0	0	0	100
B11	0	4	96	0	E11	0	4	92	4	H11	0	0	0	100
B12	0	0	0	100	E12	0	0	100	0	H12	0	0	0	100
B13	0	0	0	100	E13	0	0	100	0	H13	0	0	0	100
B14	0	0	0	100	E14	3	0	97	0	H14	0	0	0	100
B15	0	0	0	100	E15	50	0	0	5	H15	0	0	0	100
C1	0	0	100	0	F1	0	0	60	40	I1	0	0	0	100
C2	0	0	100	0	F2	2	0	98	0	I2	0	0	0	100
C3	11	37	52	0	F3	1	0	99	0	I3	0	0	0	100
C4	4	0	96	0	F4	3	1	96	0	I4	0	0	0	97
C5	18	0	82	0	F5	3	0	97	0	I5	3	0	95	2
C6	34	19	47	0	F6	42	5	53	0	I6	0	0	97	3
C7	9	1	34	36	F7	9	29	57	5	I7	0	0	100	0
C8	5	15	19	41	F8	0	4	96	0	I8	0	0	0	100
C9	3	17	56	24	F9	0	0	96	4	I9	0	0	0	100
C10	6	2	92	0	F10	3	0	94	3	I10	0	0	0	100
C11	0	5	93	2	F11	0	0	86	14	I11	0	0	0	100
C12	0	0	0	100	F12	0	10	85	5	I12	0	0	6	100
C13	0	0	0	100	F13	100	0	0	100	I13	0	0	0	100
C14	0	0	0	100	F14	0	0	100	0	I14	0	0	0	100
C15	0	0	0	100	F15	0	0	0	100	I15	0	0	0	100

Urban areas, as well as being classified according to land use, may also be differen-
tiated according to the characteristics of their residents. Shevsky and Bell (1955)
identified three dimensions of social segregation — socio-economic, family, and
ethnic. Murdie (1969) has proposed a model which incorporates all three dimensions
(Figure 3.20): the pattern of socio-economic status forms a number of sectors, the
pattern of family status a series of concentric rings, and the pattern of ethnic status
distinct nuclei. Unfortunately Murdie's model is difficult to apply in the British
context because the population is not as residentially mobile as that of the USA, and
also because local authority housing in the UK accounts for over a quarter of the
total housing stock.

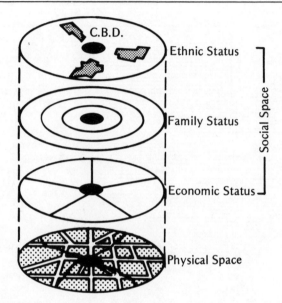

Figure 3.20   A model of the residential structure of a city
Reprinted with permission from "Factorial ecology of metropolitan Toronto 1951—61" by R. A.
Murdie (1969), Department of Geography, University of Chicago, *Research Paper No. 116*, figure 1.

RESIDENTIAL LAND USE IN PORTSMOUTH AND PRAGUE

Decision making by important individuals and corporate bodies such as the Government can have a significant influence on residential patterns. R. C. Riley (1972) has shown that, unlike many south coast resorts, Southsea owes its early expansion to the development of the Royal Naval Dockyard at Portsmouth which attracted a large officer class to the area. By the early nineteenth century the artisan Croxton Town had expanded to become middle-class Southsea, an outlier of the Dockyard and Garrison (Figures 3.21a and 3.21b). Meanwhile rapid expansion was taking place in an easterly direction from Portsea, this haphazard pattern of development resulting from fragmented land ownership which inhibited the familiar continuous urban sprawl. Between 1830 and 1860, Southsea was developed by Thomas Owen, a

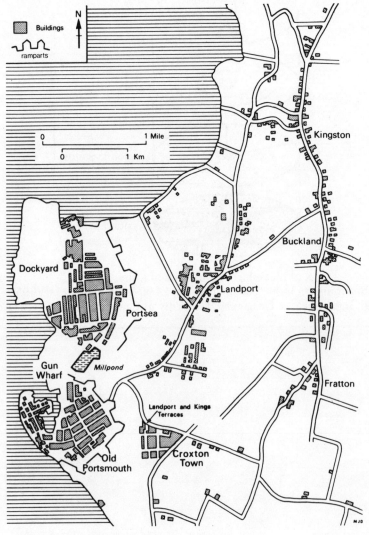

Figure 3.21a   The extent of development of Portsmouth, 1810

property speculator, architect, and builder who created this area for naval officers and the affluent citizens of Portsmouth; it thus became the most spacious and elegant part of Portsea Island and contrasted sharply with the rows of working men's houses that were being built in Fratton, Buckland, and Kingston.

The most rapid urban expansion occurred during the late nineteenth century when Southsea emerged as a holiday resort. The construction of a tramway network initiated more terraced housing across the middle of the island so that by 1914 the urbanization of Portsea island was almost complete (Figure 3.21c). Future developments of Portsmouth were to take the form of extensive new housing estates on the mainland, and redevelopment of existing built-up areas.

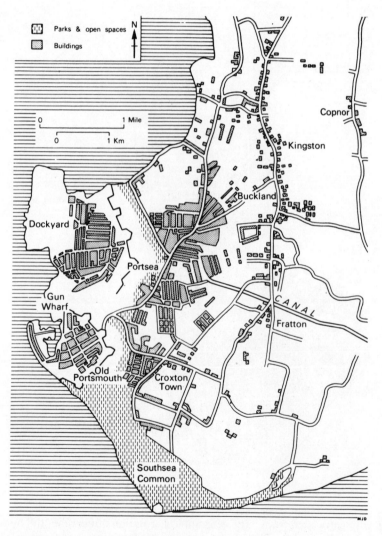

Figure 3.21b   The extent of development of Portsmouth, 1833

The Government, in addition to providing a market for Owen's Southsea, had also prevented the development of Southsea Common. The extensive area of Southsea Common was of strategic value to the Army, and even when it was sold to Portsmouth Corporation in 1923, restrictions were imposed on its development.

1   Using the information given in Figures 3.22a, b, and c, describe how far the urban structure of Portsmouth conforms to standard urban land-use models.

In contrast to the cases we have studied, the majority of land in Eastern Europe is state owned and therefore bid-rent theory cannot adequately explain the pattern of

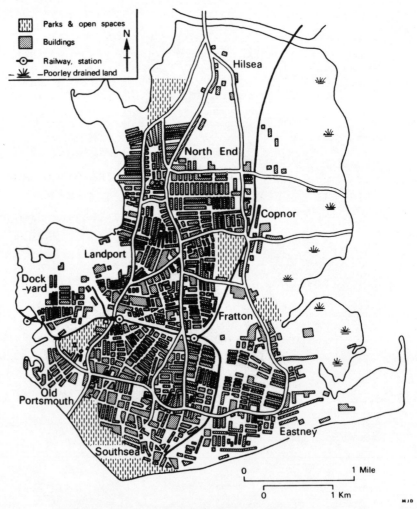

Figure 3.21c   The extent of development of Portsmouth, 1913
All reprinted with permission from "The geographical evolution of Portsmouth" by J. Chapman (1974), *Portsmouth Geographical Essays* published by the Department of Geography, Portsmouth Polytechnic, figures 2, 3 and 4.

urban land use. Social costs are considered to be more important than private costs and even the population densities are controlled to give a uniform spread of city population.

Jiri Musil (1968) has examined the ecological structure of Prague and has observed major changes as a result of Government policy. In 1930, the inner residential zone was occupied by the upper and middle classes, the middle zone by the white-collar workers, and the outer zone by the manual workers (Figure 3.23). By 1960 this pattern had been broken up by a housing policy which gave substantial rent subsidies to its tenants. This policy also did not let location influence rent and favoured "key economic" workers, young families, and those living in poor-quality dwellings.

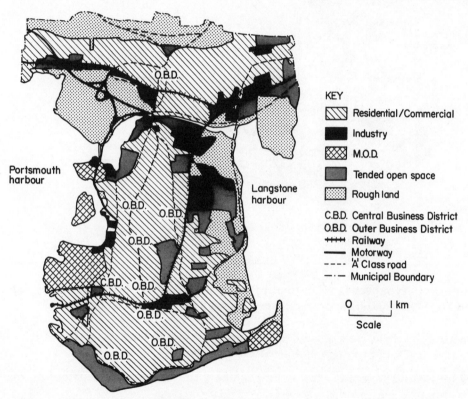

Figure 3.22a   Portsmouth: land use
Reprinted with permission from *Portsmouth Atlas* published by the Department of Geography, Portsmouth Polytechnic, part of maps 5.5 and 5.8a.

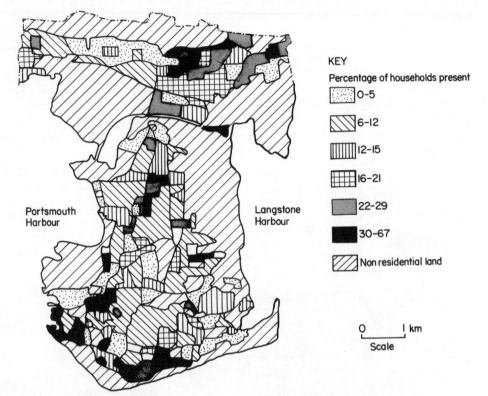

Figure 3.22b   The distribution of employers, managers, and professional workers in Portsmouth

Consequently younger families tend to occupy the modern housing and the segregated zones of 1930 have been replaced by socially mixed neighbourhoods.

2   Construct simple models of residential land use and population density which illustrate the structure of Prague in both 1930 and 1960.

3   How far is Government control in Prague likely to influence the distribution of other urban activities?

4   What is striking about the distribution of manual workers in Prague in 1930? Attempt to explain this distribution (see Figure 3.23).

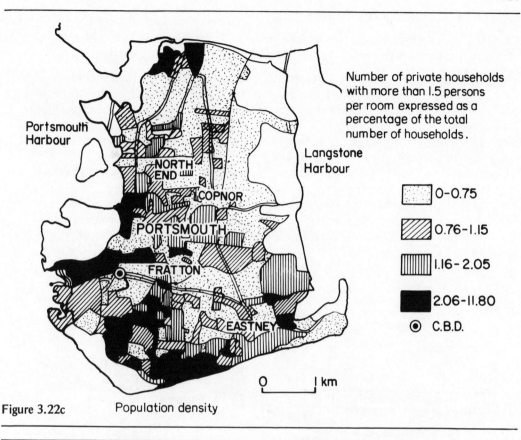

Figure 3.22c   Population density

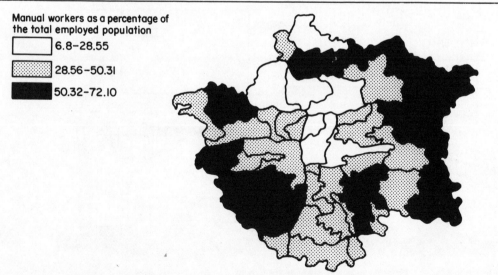

Figure 3.23   Percentages of manual workers by area in Prague, 1930
Reprinted with permission from J. Musil (1968) in *Readings in Urban Sociology* edited by R. E.
Pahl, published by Pergamon Press.

CENTRAL BUSINESS DISTRICTS

Central business districts are the heart of a city's commercial, social, and cultural life; they possess a number of clearly defined characteristics which include:

i   a high degree of accessibility — the central business district develops in and around the focus of the public transport and road systems and thus has more pedestrian traffic and, in some cities, more road traffic than any other area of the city (Figure 3.24a);

ii   high land values — the supply of accessible sites is very limited, giving rise to intense competition between urban activities;

iii a concentration of tall buildings — the multi-storey buildings afford an effective means of utilizing this scarce land intensively (see Figure 3.24b); and

iv   low residential population — the daytime population far exceeds that of the residential population, and the population density at the city centre is generally less than the next outer zone, thus forming a density crater (Figure 3.25) which broadens with time as the city expands.

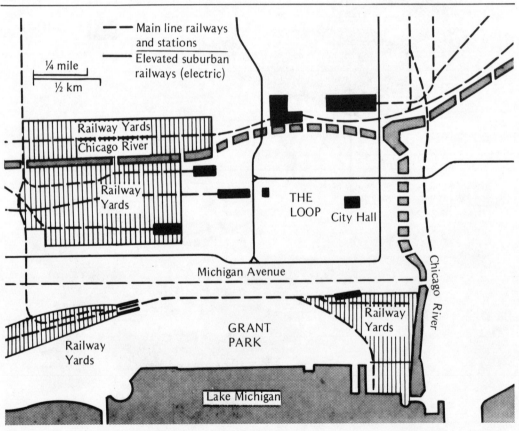

Figure 3.24a   Chicago: sketch map of the Loop

Figure 3.24b   Chicago: view northwards showing Grant Park in the foreground and high-rise ribbon development along the coast
Reproduced by permission of Aerofilms Library.

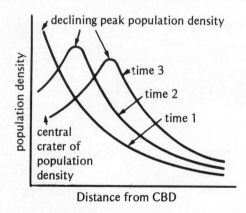

Figure 3.25   Newling's model of urban population density

164

1   With reference to Figures 3.26a, b, and c, suggest why Toronto does not conform to Newling's hypothesis illustrated in Figure 3.25.

Much analytical work on CBDs, as they are known, has been conducted by Murphy and Vance (1954), who devised the Central Business Index (CBI) technique in order to delimit these areas; a standardized technique is vital if one wishes to compare a number of CBDs and draw conclusions from them.

The CBI method requires the detailed mapping of all the land uses in a city central area in which the CBD is assumed to be located and the classifying of these uses into CBD and non-CBD categories. According to Murphy and Vance, non-CBD uses are those listed in Table 3.8, all other urban land uses being considered as CBD uses.

Table 3.8   Non-central business district uses

Permanent residences (including apartment houses and rooming houses)
Government and public (including parks and public schools, as well as establishments carrying out
    city, county, state, and federal government functions)
Organizational institutions (churches, colleges, fraternal orders, etc.)
Industrial establishments (except for newspapers)
Wholesaling
Vacant buildings and stores
Vacant lots
Commercial storage and warehousing
Railroad tracks and switching yards

Reprinted with permission, from R. E. Murphy and J. E. Vance (1954). Delimiting the CBD. *Economic Geography*, 30, 189–222. (Table II, p. 204).

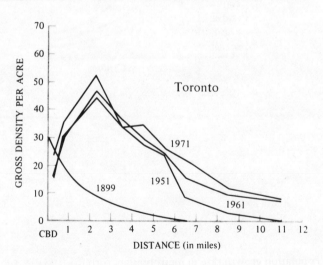

Figure 3.26a   Population density gradients in Toronto

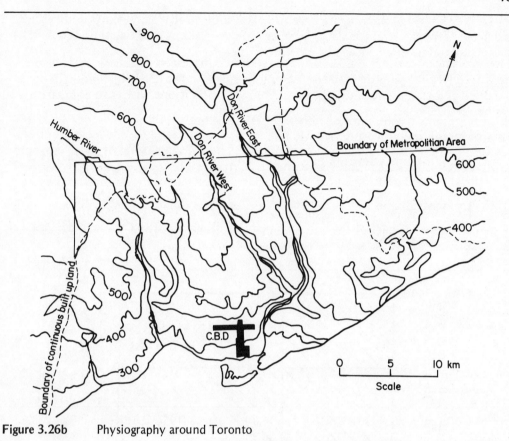

Figure 3.26b   Physiography around Toronto

Figure 3.26c   Population growth rate in metropolitan Toronto, 1951 to 1961. Isolines show percent change over the ten year period. Reprinted Fig. 10.4 (p. 230) in the North American City, second edition by Maurice Yeates and Barry Garner. Copyright © 1971, 1976 by Maurice H. Yeates and Barry J. Garner. By permission of Harper & Row, Publishers, Inc.

For each city block, which is the basic US areal unit of urban areas, two indices are calculated from the land use data; these are the central business height index (CBHI) and the central business intensity index (CBII), where:

$$CBHI = \frac{\text{Total floor area devoted to CBD uses}}{\text{Total ground floor area}}$$

which represents the number of floors of CBD use that would exist in the block if these uses were spread evenly over it; and

$$CBII = \frac{\text{Total floor area devoted to CBD uses}}{\text{Total floor area}} \times 100$$

which represents the percentage of the total floor area devoted to CBD uses.

To be considered as part of the CBD a block should have a CBHI value of at least 1.0 and a CBII value of at least 50 percent (Figure 3.27).

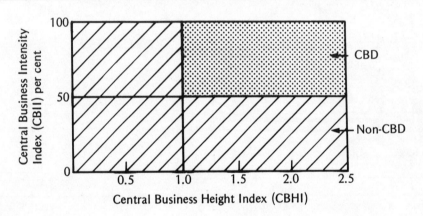

Figure 3.27

2 Refer to Figure 3.28 and approximate the boundary of Brighton's CBD. Suggest reasons for the shape and orientation of Brighton's CBD.

Malcolm Proudfoot made a detailed study of a number of American cities and identified five types of retail structure:
  i   the central business district;
  ii the outlying business centre (OBC) — similar to the CBD but smaller;
  iii the principal business thoroughfare — characterized by heavy mass and vehicular traffic between the CBD and OBC;
  iv the neighbourhood business street — whose customers walk from home; and
  v   the isolated store cluster — which usually comprises a few complementary convenience goods stores.

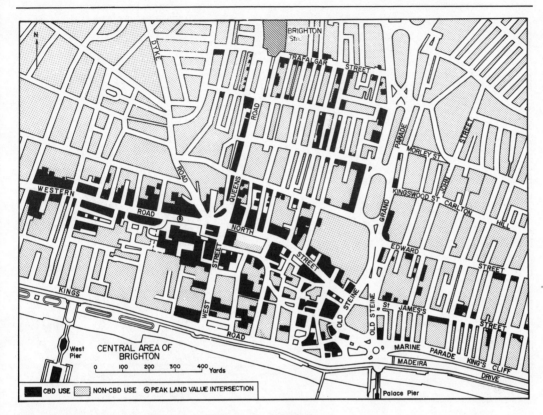

Figure 3.28 The distribution of CBD uses in Brighton

3 To what extent does the distribution of CBD activities in Brighton correspond to Proudfoot's findings given that most of Brighton's CBD "blocks" possess retailing functions on the ground floor?

4 What are the likely problems associated with applying Murphy and Vance's Central Business Index technique to British cities and resort towns?

5 Figure 3.29 indicates the census points of a pedestrian survey conducted simultaneously during a ten-minute period of normal shopping hours.

a Construct isopleths at intervals of 25 persons and comment on your results.

b How far do pedestrian density variations mirror the localization of CBD activities?

c What evidence is there of outer business district activity?

6 Central business districts lend themselves to analysis by transect.

Table 3.9 contains data on average gross rateable values and pedestrian flow for 30-metre sections (north and south sides) of Western Road, North Street, and Castle Square, a main thoroughfare which runs through the heart of Brighton's CBD.

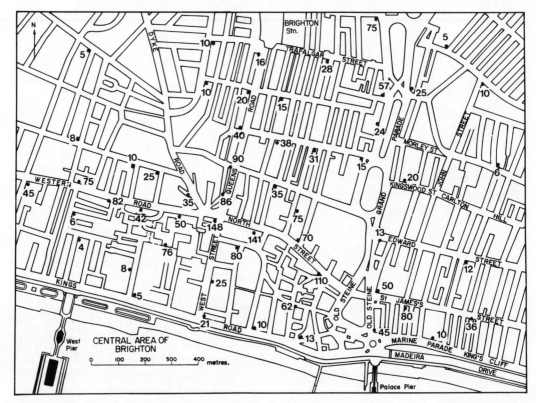

Figure 3.29   The distribution of pedestrians

Construct a composite histogram (Figure 3.30) to show the relationship between pedestrian flow and average gross rateable value.

7   Statistically test the strength of this relationship by calculating Spearman's rank coefficient of correlation and use the graph (Figure 1.25, p. 43) to test if the correlation is significant.

8   Within the CBD there is a very sharp decline in land values a short distance away from the peak land value intersection and it is greatest along the latitudinal axis.

With reference to Table 3.9, plot gross rateable values as a scattergraph (Figure 3.31) and from the PLVI construct lines of best fit or regression lines (Figure 3.32). With what percentage of the PLVI does the boundary you have drawn on Figure 3.30 appear to correspond?

Table 3.9

	Gross rateable value (£)	Pedestrian flow (persons per 30 minutes)
West	213	44
	310	45
	379	55
	575	57
	1027	60
	950	35
	586	46
	690	73
	1580	85
	1975	103
	1430	105
	1490	160
	1205	150
	2400	152
	1413	113
*P.L.V.I. →	3250	126
	2330	130
	1400	88
	600	69
	804	55
	691	68
	970	87
	1220	116
	835	105
	592	97
	820	86
	785	70
	1043	94
	600	68
East	220	66

*Peak land value intersection — so called because the highest
land value generally occurs near an important road intersection
in the American urban grid-iron pattern.

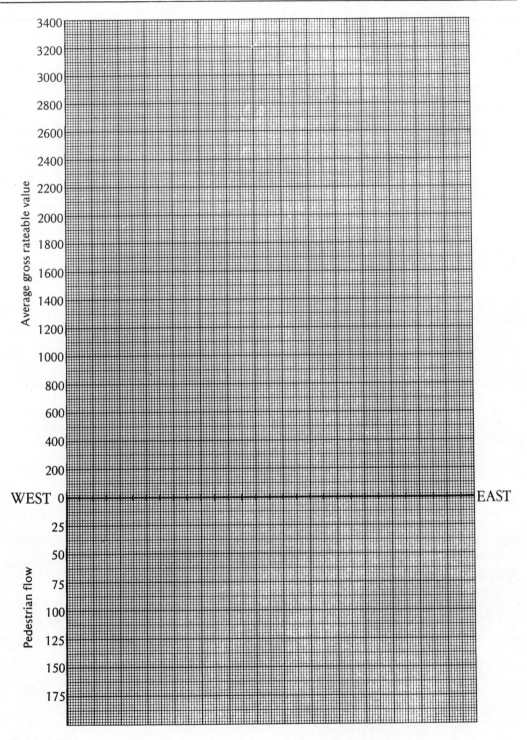

Figure 3.30

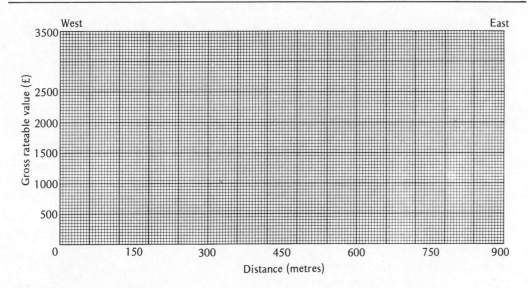

Figure 3.31

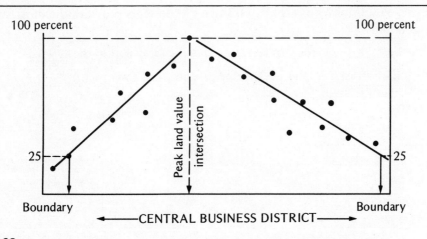

Figure 3.32

PLANNING THE STRUCTURE OF A NEW TOWN

In an earlier exercise (p. 74) you sited a new town for 75,000 people in the Tickhill area. To accommodate a total population of 75,000, an area of approximately 16 square kilometres would have to be built over.

1 On the sketch map (Figure 1.40) draw a kilometre grid and select 16 grid squares (or parts of grid squares) which you think will provide the most suitable area for the development of the new town. These need not form a regular shape.

Complete on enlarged version of the grid map (Figure 3.33) by locating and labelling the following features:

    i   the Central Business District
   ii   the primary road system
  iii   open spaces
  iv   the industrial sector(s)
   v   primary and secondary schools
  vi   neighbourhood units and their shopping centres
 vii   a multi-purpose entertainment and recreation centre
viii   hospital

2 Having completed the map:
 a Explain why you think your area of 16 square kilometres provides the most suitable area for development.

 b Discuss the decisions you took in selecting specific sites within the town.

3 Compare and contrast the land use zoning of your new town with:
   i   an existing new town
  ii   your local area

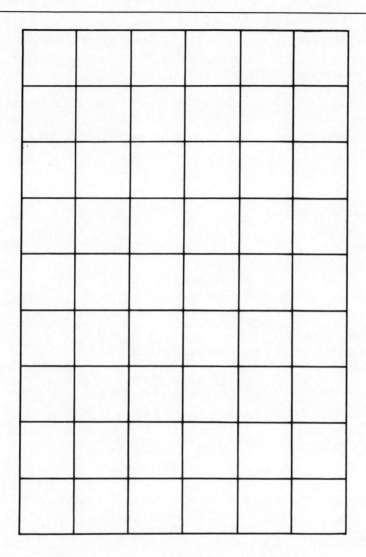

Figure 3.33

Further Reading

The Geography of Retailing, P. Scott, Hutchinson University Library (1970).

The North American City, M. H. Yeates and B. J. Garner, Harper and Row (1976), Chapters 9, 10, 11, and 12.

Inside the City, J. Everson and B. FitzGerald, Longman (1972).

Human Geography: Theories and Their Application, M. G. Bradford and W. A. Kent, Oxford University Press (1977), Chapter 5.

Urban and Regional Planning, P. Hall, Pelican (1974).

CHAPTER FOUR
THE ENVIRONMENTAL IMPACT
OF SETTLEMENTS

Settlements are composed of houses, commercial and industrial buildings, streets, parks, and so forth; they thus form an appreciable interruption to the rural landscape and can modify both the climate and the flood characteristics and water quality of rivers in the vicinity of the area they occupy.

The Climatic Impact

Settlements alter climate by influencing the spatial pattern of temperature, water content, and pollutant content of the near-surface atmosphere. In the countryside, some rainfall is intercepted by vegetation, some of it is soaked up by the soil, and some is passed back to the atmosphere by transpiring plants. In an urban area, impermeable surfaces, rapid removal of surface run-off through drains and gutters, and the reduction in transpiration resulting from the relatively sparse vegetation cover, usually lead to a lower water content in an urban area compared with the rural surrounds. The earth's surface energy budget is also modified by settlements since in the urban area there is usually less water to evaporate from the ground than in the rural area and so more energy is available to heat the ground, the urban fabric, and the overlying air. It is not surprising therefore that, at least for windspeeds not exceeding seven metres per second, temperatures over urban areas are usually higher than over adjacent rural areas; this urban heat island phenomenon was first observed by Luke Howard for London in 1833.

178

Another reason for the urban area being warmer than the countryside is the large heat capacity of the urban fabric, which by far compensates for its high thermal conductivity: an object with a high heat capacity can store much heat, and an object with high thermal conductivity heats up and cools down rapidly. So, during daylight hours, settlements can build up a large store of heat which is released into the urban air during the night.

Of the other factors which might lead to rural-urban differences of the kind mentioned, the reflectivity or albedo of surface materials and the configuration of the surface deserve attention. Clearly, the more incoming solar radiation a surface reflects, the less will be the energy absorbed by the surface. Concrete when dry has an albedo which is approximately 10 percent greater than that of woodland or grass; it therefore reflects more solar radiation. Tarmac roads, however, have an albedo which is some 5 percent lower than that of woodland and grass. Consequently, the influence of rural-urban albedo differences is likely to be minimal except perhaps on a local scale where "hot spots" may develop over tarmac. The configuration of the surface in the urban area can lead to a near-surface gain or loss of heat. Air in streets, on the one hand, may be trapped and heated by long-wave radiation emitted by buildings; on the other hand, the "roughness" of the buildings tends to break up the air-flow and increase turbulence which in turn encourages the upwards transfer of warm air and hence the cooling of the urban air.

Settlements use fuels for residential, industrial, commercial, and transport purposes. The combustion of fuels is concentrated in a small area in most settlements and the overlying air may become contaminated with atmospheric pollutants (Figure 4.2). The most prevalent urban air pollutants are smoke particles and sulphur dioxide (Figure 4.1).

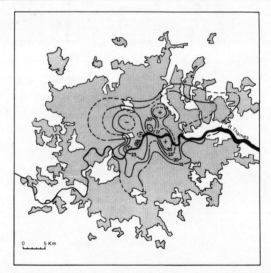

Figure 4.1   The average distribution of smoke in London from April 1957 to March 1958. Units are mg/100m^3
Reprinted with permission from *The Climate of London* by T. J. Chandler (1965) published by Hutchinson.

Figure 4.2   Smog formation over Edinburgh.
Smog density increases over industrial Leith (right) and the smog merges with sea mist over Firth of Forth (top right)
Reproduced by permission of Aerofilms Library.

Combustion also produces heat which may boost rural-urban temperature differences. It has been estimated that large German cities produce between 15 and 30 calories/ cm^2 /day, which compares with direct insolation (solar energy) of 52 and 518 calories/cm^2 /day in December and June respectively.

THE STUDY OF A SCHOOL'S HEAT ISLAND

For the schools shown in Figure 4.3, meteorological data were obtained for 35 uniformly spaced reference points in the school grounds using a whirling psychrometer, an instrument which gives wet and dry bulb temperatures from which relative humidity may be calculated. In addition to noting the prevailing weather conditions, the relevant synoptic chart was consulted for the study period (a.m., 11 May, 1976). The urban heat island effect is likely to be most pronounced under relatively stable conditions. During the period of this survey the area lay within a narrow ridge of high pressure which was sandwiched between two frontal systems, one lying over mainland Europe, the other lying over Ireland. The weather was dry with sunny intervals and a light (1—2 knots) southwesterly wind.

1   With the aid of Table 4.1* draw isopleths at intervals of 2 percent, from 56 to 66 percent, on the map provided (Figure 4.3) to show the variation in relative humidity throughout the school grounds.

2   Briefly describe and account for the pattern revealed by the isopleths. Given the linear layout of the school buildings, can you suggest why the pattern of isopleths is not symmetrical?

3   For a sample of points (Nos. 8 to 28 inclusive) test the influence of the school's urban fabric upon temperature by calculating Spearman's rank correlation coefficient using effective urban area and temperature data from Table 4.1. Test the statistical significance of your result by referring to Figure 1.25 (p. 43).

4   Suggest, and give reasons for, two meteorological conditions that would make the heat island effect (that is, the difference between urban and rural temperatures), most marked during the day and night.

5   What would be the main problems of conducting a similar study on a city scale?

---

*
The data were obtained by using a whirling psychrometer and recording the wet and dry bulb temperatures, from which relative humidity is calculated, at 35 uniformly spaced reference points. The weather conditions at that time (a.m., 11 May, 1976) were fair and dry with a light (force 2) southwesterly wind. The percentage of urban area was calculated by placing a net of 35 grid squares over the map, each square being centred on reference points. The data therefore refers to the percentage of the area composed of buildings and extensive tarmac and recreational surfaces.

Table 4.1

Ref. No.	Relative Humidity (%)	Dry Bulb Temp. (°C)	Urban Area (%)
1	60	13.0	0
2	61	13.1	0
3	61	13.1	0
4	61	13.1	0
5	62	13.0	0
6	65	12.2	0
7	66	12.3	0
8	56	14.4	36
9	60	13.8	1
10	60	13.8	3
11	59	13.5	24
12	60	13.9	0
13	64	12.5	0
14	66	12.2	0
15	55	15.5	22
16	57	14.3	80
17	57	14.6	92
18	58	14.6	5
19	59	14.8	90
20	60	13.8	27
21	65	13.1	0
22	60	13.8	0
23	58	13.5	6
24	59	13.6	2
25	59	13.6	2
26	60	13.8	9
27	61	14.0	3
28	64	13.5	0
29	61	13.2	0
30	61	13.2	0
31	61	13.4	0
32	61	13.2	0
33	61	13.2	0
34	62	13.5	0
35	64	13.5	0

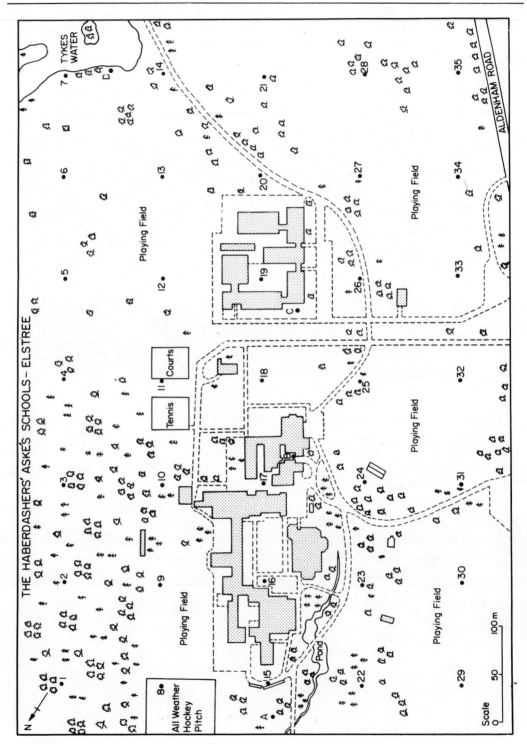

Figure 4.3

On the graph paper (Figure 4.4), suggest a diurnal pattern of temperature variations that you would find in a rural and urban area during the summer by completing the three curves. Explain the pattern you have predicted.

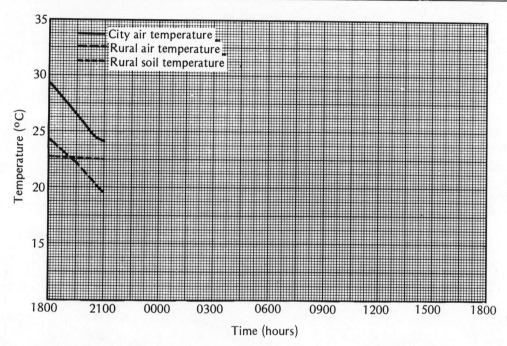

Figure 4.4   Rural and urban surface temperatures

Recent work has shown that a relationship exists between the size of urban areas and the intensity of heat islands.

With the aid of Table 4.2, construct a scattergraph and draw a best-fit line to show the strength of the correlation between observed maximum urban heat island temperatures (the difference between maximum city and rural temperatures) and population size for selected European cities.

Table 4.2

	Magnitude of heat island (°C)	Population
London	10.0	7,168,000
Berlin	10.2	3,137,000
Vienna	8.1	1,859,000
Munich	7.1	1,337,000
Sheffield	8.0	507,000
Utrecht	6.0	463,000
Malmö	7.4	451,000
Karlsruhe	7.1	261,000
Reading	4.2	133,000
Uppsala	6.5	136,000
Lund	5.7	35,000

Source: Based on T. Oke (1973)

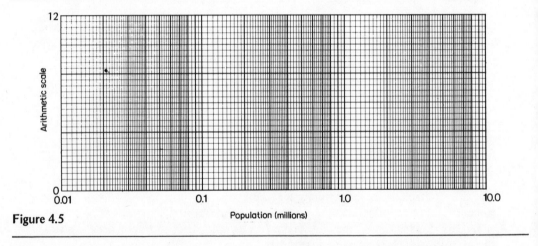

Figure 4.5

8 Given that data for North American cities produces a steeper gradient, what does this suggest about the difference between American and European heat islands?

9 As heat islands can cause sweltering summer nights it is desirable to plan cities so that the heat island effect is minimized. From all the relationships you have considered so far, suggest possible ways of reducing the heat island effect.

Heat islands also have significant depth: they are domes of warmed air. In situations where temperature inversions are common, the presence of a large industrial city introduces a further complexity into the vertical pattern of air temperatures.

With the aid of Table 4.3, construct a composite graph (Figure 4.6) to show the vertical variation of air temperatures
 i above Cincinnati city centre
ii above its rural environs — 20 kilometres from the city centre.

Mark by a horizontal line and label the convective lid (temperature inversion).

Explain the pattern of the following temperature gradients

  i Rural environs 0 to 225 metres
 ii Rural environs 225 to 500 metres
iii Cincinnati 0 to 50 metres.

What are temperature inversions and how do they aggravate the problem of air pollution? Illustrate your answer with reference to one named city, for example, Los Angeles.

Table 4.3

Height above ground level (metres)	Downtown Cincinnati air temperature (°C)	Rural environs air temperature (°C)
0	21.7	18.0
25	20.5	18.3
50	20.0	19.0
75	20.4	20.1
100	21.2	21.1
125	22.0	22.0
150	22.7	22.7
175	23.5	23.5
200	24.4	24.4
225	25.0	25.0
250	24.9	24.9
275	24.7	24.7
300	24.5	24.5
325	24.3	24.3
350	24.1	24.1
375	23.9	23.9
400	23.7	23.7
425	23.5	23.5
450	23.3	23.3
475	23.1	23.1
500	22.9	22.9

Source: Adapted from work by J. F. Clarke (1969)

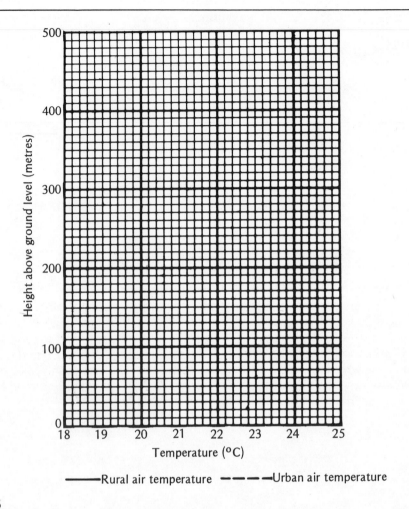

Figure 4.6

In summary, settlements create their own climates: compared with the surrounding countryside, settlements, at least the large ones, have fewer days of snow, a longer growing season, lower heating bills, but also more polluted air, poorer visibility, less sunshine, and more rainfall and fog.

The Hydrological Impact

Settlements may also modify the hydrology of catchments they occupy. Relatively impermeable urban surfaces — tiles and concrete — as well as gutters, drains, and sewerage systems mean that rainfall is routed much more quickly to streams than it would be in an adjacent rural catchment; this leads to changes in the flood characteristics of rivers which are fed by an urban catchment. Generally, the changes induced in river flow are unwanted: floods occur more frequently, and tend to be larger, than in a similar rural catchment.

FLOODING IN NORTH WEST LONDON*

Cities are particularly vulnerable to flood hazard since high-intensity rainfall results in very rapid run-off into a limited number of artificially contained river channels which can easily overtop their banks. Such a situation occurred on the night of the 16th to 17th August 1977 in the River Brent catchment and caused considerable damage to private and commercial property as well as disrupting communications, as the following extract from *The Times* newspaper, and Figure 4.7b, show.

The Times Thursday August 18 1977

Floods disrupt travel in the South

By Hugh Clayton

Travel disruption. The storms disrupted road and rail travel and telephone services in southern England and the Midlands (the Press Association reports). Almost 24 hours later, floodwater was still causing difficulties in north and north-west London. Some roads in Greenford were still under 6ft of water, and diversions were set up where the Grand Union Canal overflowed on to the North Circular Road. Some cars on the road were submerged.

Many rivers, including the Thames, were still high last night, and there were fears that further rain would add to the disruption.

The police evacuated more than 30 people from homes in the Greenford area.

Among at least 20 main London roads badly affected by the flooding were Chelsea Embankment, Brent Cross, and Hanger Lane at Ealing.

Train services to and from Euston were subject to delays. Local services from Bedford, St. Albans and Luton, which normally run to Moorgate, were diverted to St. Pancras, and other services north of London were disrupted.

Many Underground stations were out of action and commuters delayed.

Nearly half an inch of rain fell at Heathrow airport last night. At Hayes, several families were evacuated in boats from their homes. They were sent to a school for the night.

The London fire service said it received so many calls for help that it had started to lose count. In the Acton area floods were up to 5ft deep. Abandoned vehicles added to the chaos.

About 70,000 telephones in London were put out of order by flooding. A restricted service operated in north-west and west London and a serious cable fault affected calls within central London.

The London Weather Centre said it was the wettest 12-hour period since August, 1971. It reported a rainfall of 1.52in on Tuesday night.

The highest rainfall reports on Tuesday night were 4.51in at Ruislip, Manor Park, and 4.44in at Maple Lodge sewage works, Rickmansworth.

Services were not running on the Underground's Central Line between White City and Queensway last night because a tunnel was waterlogged. British Rail trains between Euston and Watford could not run between Willesden Junction and Harlesden. Passengers had to use buses.

Reproduced from
The Times by permission.

---

* This exercise is based on an internal assessment unit devised by Michael G. Day for the Schools' Council Geography 14–18 Project (Bristol).

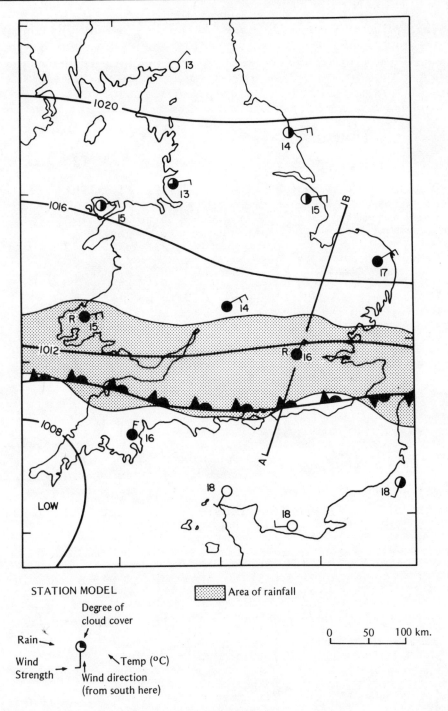

STATION MODEL   ▒▒▒ Area of rainfall

Degree of
cloud cover

Rain →   ☽

Wind →
Strength   ↑
          Wind direction
          (from south here)

↖ Temp (°C)

0   50   100 km.

Figure 4.8   Weather map for 00 hrs, 17 August 1977

Figure 4.7a   Pinner Station "normal" state

1   Figure 4.8 shows that the cause of serious flooding in northwest London on the
     night of the 16th to 17th August 1977 was the movement of an occluded front
     across the city in a northwesterly direction.

     With reference to Figure 4.8, mark and label the following on Figure 4.9

       i   warm front
      ii   cold front
     iii   line of occlusion
      iv   warm sector aloft
       v   area of rainfall on the ground
      vi   temperature before and after occluded front
     vii   general position of London
    viii   general position of south and east coasts

Figure 4.7b   Morning commuters brave the water hazard on the way to Pinner station
Photograph G. G. Attridge

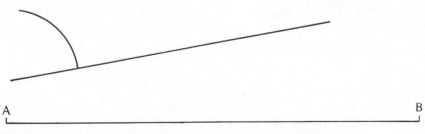

A ├─────────────────────────────────┤ B

Figure 4.9   A diagrammatical cross section through an occluded front

You have all seen a stream and noticed that the amount of water moving down it varies considerably with time. Sometimes the water level is low in the channel, or the bed may even be dry; at other times, during flood, the stream may overflow its banks. Water moving down a stream is known as streamflow. The amount of stream-flow passing a given cross section of a stream per unit time is stream discharge; it is usually denoted by the letter Q and measured in cubic metres per second (m^3/sec or cumecs). A hydrograph of a river is a graph which shows how the streamflow (discharge) varies with time. As such, it reflects those characteristics of the catchment which influence run-off. A hydrograph may show yearly, monthly, daily, or instantaneous discharges and from it may be determined base flow, which is more or less the dry weather discharge of the stream, and periods of high flows or flood discharges. The total flow is base flow plus flood flow. The components of a hydrograph are shown in Figure 4.10a, in which rainfall occurs over a five-hour period and represents a single storm event. The discharge of the stream changes in response to the rainfall input to the catchment but, since most of the rain water has to enter the soil and move through or over valleyside slopes before it reaches the stream, the response in the stream discharge is not immediate.

Two extreme types of hydrograph reflect extremes of streamflow response to rainfall (Figure 4.10b). Firstly, a flashy response occurs when the rainfall gets quickly into the stream and the hydrograph is characterized by steep limbs and a sharp peak; secondly, a steady response occurs when the rainfall seeps slowly to the stream and the hydrograph is attenuated and smooth. Since the shape of the hydrograph depends upon the features of the catchment that determine the disposition of rainfall, its interpretation is an important tool in the analysis of stream and basin characteristics. The rising limb of the curve is generally concave upwards and reflects the infiltration capacity of the catchment; the time before the steep climb represents the time before infiltration capacity is reached in those areas where overland flow is likely. A sudden, steep rising limb reflects great immediate surface runoff and little infiltration; this may be owing to steep, short slopes, the narrowness of the catchment, thin soils, poor vegetation cover, or a high percentage of impermeable surface. The crest, or peak, of the curve marks the maximum runoff; with the same storm, a catchment with large soil storage capacity has a lower peak than one with little storage capacity. The recession curve or receding limb represents the outflow from the catchment after inflow has ceased; its slope depends on the physical characteristics that determine storage.

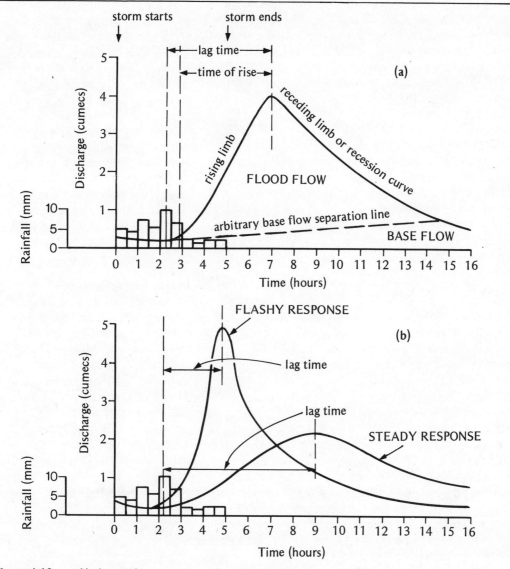

storm starts   storm ends

Figure 4.10   Hydrographs

Using data from Table 4.4:

a Construct a bar graph on Figure 4.11 to show rainfall at 3-hour intervals over the period 1800 hrs 16 August to 0900 hrs 18 August.

b Construct a graph on Figure 4.11 to show discharge at Wealdstone Brook.

c Calculate the lag-time for each of the three data stations.

Table 4.4

Time		Rainfall Total at end of 3 hr. period (mm)	Discharge Height above crest of measuring weir (metres)		
			Wealdstone Brook	Monks Park	Hanwell
16/8/77	18.00	0	0.1	0.4	0.5
	21.00	0	0.1	0.4	0.5
17/8/77	00.00	19	0.5	1.2	0.7
	03.00	27	1.1	1.8	1.8
	06.00	23	1.3	2.5	2.3
	09.00	0	1.0	2.8	3.0
	12.00	0	0.5	2.6	3.3
	15.00	0	0.3	1.2	3.0
	18.00	0	0.25	0.7	2.5
	21.00	0	0.25	0.6	1.5
18/8/77	00.00	0	0.25	0.6	1.0
	03.00	0	0.2	0.6	0.9
	06.00	0	0.2	0.6	0.8
	09.00	0	0.2	0.1	0.8

2 Examine Figure 4.12 and consider very carefully the position of the three gauging stations.

a Describe and explain the shape of the hydrographs.

b Why do you think the lag-time increases in a downstream direction?

c Why do you think that the Wealdstone Brook gauging station recorded the least peak discharge?

d What effect do you think the Brent Reservoir may have had on the response of the river downstream to the heavy rainfall?

e Discuss the factors that would cause a similar stream response in a rural area.

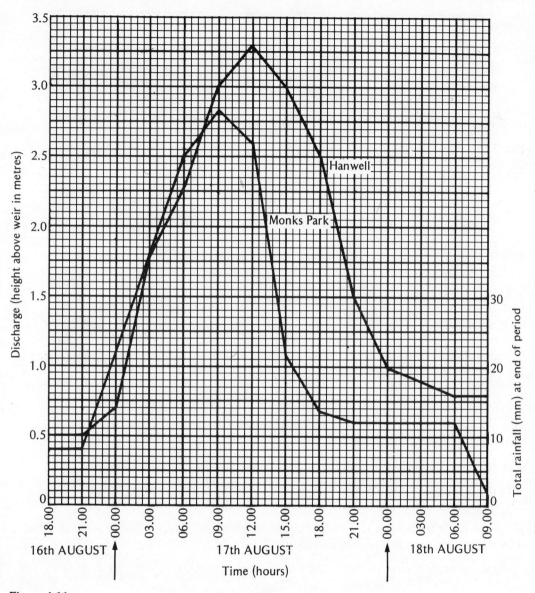

Figure 4.11

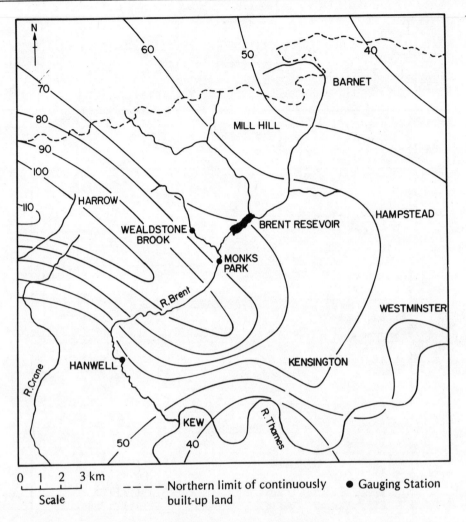

Values show total rainfall in mm.

Figure 4.12   Isohyetal map for 16 August 1977, 24 hrs ending 09.00 on 17 August. Values show total rainfall in millimetres
Source: D. Ayres F.I.C.E., F.I.Mech.E., F.I.P.H.E., Director of Public Health Engineering, Greater London Council.

THE EFFECT OF BUILDING HARLOW NEW TOWN ON FLOODING IN CANON'S BROOK

This exercise is concerned with the effect of urbanization on streamflow, and especially floodflow, characteristics. To this end, data on streamflow in Canon's Brook catchments, Harlow, Essex, from 1952 to 1968 will be used. The task will be to study the data and answer questions which refer to the changing nature of stream-flow characteristics which have accompanied the progressive urbanization of the catchment (see Figure 4.13).

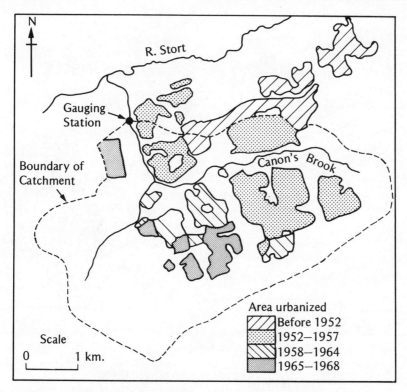

Figure 4.13 The urbanization of Harlow

Reprinted with permission from "The effect of urbanization on floods in Canon's Brook, Harlow, Essex" by G. E. Hollis (1974). In *Fluvial Processes in Instrumented Watersheds* edited by K. J. Gregory and D. E. Walling, IBG Special Publication No. 6, simplified version of figure 1.

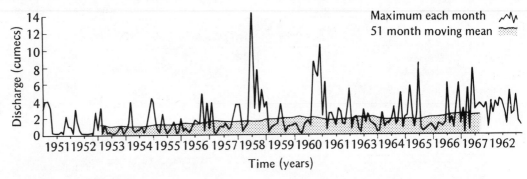

Figure 4.14   Discharge in Canon's Brook
Reprinted with permission from "The effect of urbanization on floods in Canon's Brook, Harlow, Essex" by G. E. Hollis (1974). In *Fluvial Processes in Instrumental Watersheds* edited by K. J. Gregory and D. E. Walling, IBG Special Publication No. 6, figure 2a.

1   Figure 4.14 shows the maximum flood each month over the period of study, as well as the 51-month moving mean for the same data. A 51-month moving mean is constructed in the following manner. The values of the floods for the first 51 months of study are added and the total divided by 51 to give the mean value for that period of time. Next, the values for months 2 to 52 would be taken and averaged in the same way; then the values for months 3 to 53 and so on. The first 51-month mean value is plotted on a graph at the 26th month (see Figure 4.14), the second value (for months 2 to 52) is plotted at the 27th month and so on. This is why on the graph the values of the 51-month moving mean start and end 26 months after and before the beginning and end respectively of the period of study.

a What do you think the 51-month moving mean does to the data?

b What pattern, if any, do you think the data show?

c Which period of urbanization appears to have had the greatest influence on the Canon's Brook floodflow characteristics? Does this correspond to a major increase in the amount of built-up land within the catchment?

2   Using data from Table 4.5, construct a composite histogram (Figure 4.15) to show the monthly variation in both rainfall and potential evaporation. Shade in different colours the areas where

   i   rainfall exceeds potential evaporation
   ii   potential evaporation exceeds rainfall

Assuming that all other factors are constant, in which season are floods most likely to occur?

Table 4.5

	Jan	Feb	Mar	Apr	May	Jun	Jul	Aug	Sep	Oct	Nov	Dec
Rainfall (mm)	61	42	40	52	46	42	64	58	54	63	67	56
Potential evaporation (mm)	1	10	33	57	84	98	98	83	52	23	5	−1

Rainfall is average (1916–1950) for North Weald (Grid Ref TL 484 063)
Potential evaporation is average for the county of Essex.

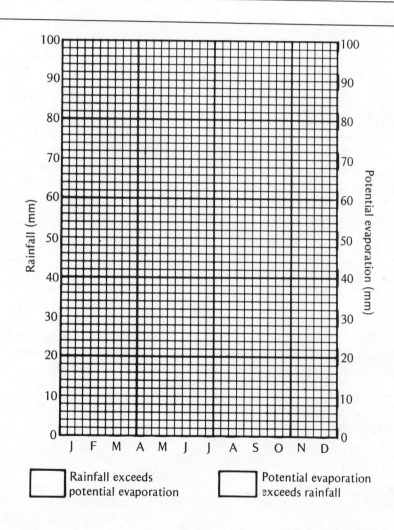

Rainfall exceeds potential evaporation   Potential evaporation exceeds rainfall

Figure 4.15

3 Figure 4.16 shows, for both the winter (October to March) and summer (April to September) months, the frequency of occurrence of three types of flood

i   floods of between 1.13 and 2.83 cumecs
ii floods of over 0.28 cumecs
iii floods of over 5.66 cumecs

Describe and attempt to account for the changes in the various types of flood in both winter and summer periods.

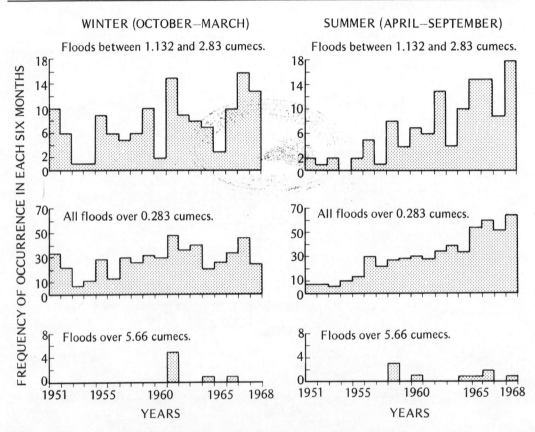

Figure 4.16   The frequency of floods in Canon's Brook, 1950 to 1968
Reprinted with permission from "The effect of urbanization on floods in Canon's Brook, Harlow, Essex" by G. E. Hollis (1974). In *Fluvial Processes in Instrumented Watersheds* edited by K. J. Gregory and D. E. Walling, IBG Special Publication No. 6, figure 2b.

Cities also produce considerable quantities of waste, both sewage and industrial effluent, much of which ends up in rivers where it changes water quality; this will be examined in the next book.

Further Reading

Geography: A Modern Synthesis, P. Haggett, Harper and Row (1975), Chapter 7.

Atmosphere, Weather and Climate, R. G. Barry and R. J. Chorley, Methuen (1976), Chapter 7. Chapter 7.

Principles of Applied Climatology, K. Smith, McGraw-Hill (1975), Chapter 3.

Run-off Processes and Stream-flow Modelling, D. Weyman, Oxford University Press (1975).

Flooding and Flood Hazard in the United Kingdom, M. Newson, Oxford University Press (1975).

Techniques in Physical Geography, J. Hanwell and M. Newson, Macmillan (1973), Chapter 5.